Creo Simulate 6.0 Tutorial

Roger Toogood, Ph.D., P. Eng.

SDC Publications

SDC Publications
P.O. Box 1334
Mission, KS 66222
913-262-2664
www.SDCpublications.com
Publisher: Stephen Schroff

ISBN-13: 978-1-63057-296-9
ISBN-10: 1-63057-296-9

Printed and bound in the United States of America.

Creo Simulate 6.0 Tutorial
(Structure / Thermal)

Preface

In his excellent text <u>Finite Element Procedures</u>, K.J. Bathe identifies two possible and different objectives for studying Finite Element Analysis (FEA) and methods: to learn the proper use of the method for solving complex but practical problems (the practitioner's goal), and to understand the methods themselves in depth so as to pursue further development of the theory (the researcher's goal). This tutorial was created with the former objective in mind. This is, by itself, a formidable task and not one that can be totally accomplished in a single, short volume. Thus, the primary purpose of the tutorial is to introduce new users to Creo Simulate and see how it can be used to analyze a variety of problems. It is assumed that this is the reader's first exposure to FEA.

The tutorial lessons cover the major concepts and frequently used commands required to progress from a novice to an intermediate user level. The commands are presented in a click-by-click manner using simple examples and exercises that illustrate a broad range of the software features and analysis types that can be performed. In addition to showing the command usage, the text will explain why certain commands are being used and, where appropriate, the relation of commands to the overall FEA philosophy.

Running Creo Simulate involves simple procedures. With a little practice (and careful judgement!) and learning only a fraction of the capabilities of the program, you can perform FEA of reasonably complex problems. This manual is meant to guide you through the major features of the software and how to use it. It is not meant to be a comprehensive guide to either the software or FEA modeling - consider it the primary school of practical FEA! Considerable experience, in particular in the critical examination of results, will be necessary to become a skilled user.

While Creo is inherently a 3D solid modeler, many FEA model types in addition to the default solid model type can be treated. These include 2D models for plane stress/strain problems and those using axisymmetric solids and shells. Shell and beam idealizations are also possible. Other enhancements to the program include cyclic constraints, changes in the interface and operation of the program such as the inclusion of simulation features in the model tree. These are all covered in this tutorial. The capability in Creo Simulate to handle large deformation problems, that is problems involving geometric non-linearity, has not been covered due to space constraints. In any case, this functionality is of a significantly more advanced level than what is required (perhaps) in an introductory tutorial. Other functionality, such as dynamic load and fatigue analysis, and treatment of hyperelastic materials, has also unfortunately been excluded.

Many aspects of the program are covered in the Creo Parametric Tutorial (also from SDC Publications) and are not repeated here (for example: location and operation of the model tree in the Navigator window, view control and the use of mouse short cuts). Because this book deals exclusively with Simulate, familiarity with Parametric is assumed.

Students with a broad range of backgrounds should be able to use this book. Some familiarity with elementary strength of materials is necessary. Because the emphasis is on Simulate, the models are not too complex and should be easily created by novice users of Creo Parametric. A number of the more complicated exercise models can be downloaded from the SDC web site located at

http://www.sdcpublications.com/Downloads/978-1-63057-296-9

This book is **NOT** a complete reference manual for Creo Simulate. There are several thousand pages available on-line with the Creo installation, with good search tools and cross-referencing to allow users to find relevant material quickly.

The tutorial treats solid models first, as these are the default model type. This is followed later by model idealizations: plane stress, plane strain, shells, beams, frames, and axisymmetric shells and solids, springs, masses, and so on. A final chapter introduces the tools for solving simple steady and transient thermal problems and for creating temperature loads used to compute thermally induced stresses.

It continues to be a challenge to decide what to include and what to exclude in this introduction in terms of the command set within Simulate. The author can only hope that the presented material will be found useful, and in the right dose! It has also been interesting to design suitable demonstration problems and exercises that are interesting, feasible with the state of learning of the user, physically meaningful, and illustrative of a broad set of Simulate functionality - all within the space of 250 or so pages. It is hoped that at least some of these goals have been satisfied.

The examples included here were executed using Creo 6.0 (version 6.0.0.0). Some users may note that with different builds, and sometimes depending on how the model was created, slightly different results may be obtained. This may be observed in the number of automatically generated elements in the mesh, iterations to convergence, maximum reported stress values, and so on. Variations from the values reported here should be no more than a few percent and will generally be within the error tolerance of the analysis.

Although every effort has been made in proofreading the text, it is inevitable that errors will appear. The author takes full responsibility for these and hopes they will not impede your progress through the tutorial. Any comments, constructive criticisms, and/or suggestions will be gratefully received and acknowledged. You can reach the author by email at *<roger.toogood@ualberta.ca>*.

Enjoy the book!

Notes to Instructors:

The tutorials consist of the following:

> 2 lessons on general introductory material (reading only, but important!)
> 2 lessons introducing the basic FEA operations using solid models
> 4 lessons on model idealizations (shells, beams and frames, plane stress, etc)
> 1 lesson on miscellaneous topics
> 1 lesson on steady and transient thermal analysis

There are very few, and all quite minor, changes from Creo 5.0 to 6.0, at least as far as the material covered in this Tutorial is concerned. There are a couple of changes in the interface (the pop-up, context sensitive menu that appears beside the cursor when anything is selected in the graphics window being one). There are no significant changes in the operation of Creo Simulate.

Each of these tutorial chapters will take between 1-1/2 to 3 hours to complete depending on the ability and background of the student. Moreover, additional time would be beneficial for experimentation and additional exploration of the program. Most of the material can be done by the students on their own; however, there are a few "tricky" bits in some of the lessons. Therefore, it is important to have experienced and knowledgeable teaching assistants available (preferably right in the computer lab) who can answer special questions and especially bail out students who get into trouble. Most common causes of confusion are due to not completing the lessons or digesting the material. This is not surprising given the volume of new information or the lack of time in students' schedules. However, I have found that most student questions are answered within the text if it is studied carefully.

In addition to the tutorials, it is presumed that some class time will be used for discussion of some of the broader issues of FEA, such as the treatment of constraints. Furthermore, it is vitally important for students to compare their FEA results with other possible solutions. This can be accomplished using simple problems for which either analytical solutions or experimental data exist[1]. An extended discussion and exploration of modeling of boundary conditions would be very beneficial, particularly in cases (I call them "diabolical") where seemingly reasonable modeling alternatives can produce significantly different results. It takes a while for some students to realize that just creating the model and producing pretty pictures is not sufficient for professional work, and the notions of accuracy and convergence need careful treatment and discussion. On the other hand, examining case after case of diabolical models that produce dubious answers can have the unintended effect of turning students off the method entirely.

It should be expected that most students, after having gone through a lesson only once, will not have "internalized" very much. My experience is that some students execute the commands without carefully reading or studying the accompanying text explaining *why*.

---

[1] See the discussion of the **Verification Guide** in the **Conclusions** section of Lesson #10.

A significant number of students are able to pick up much of the program by trial and error[2]. The second pass through the lesson usually results in considerably more retention and broader understanding. Each lesson concludes with a number of review questions and simple (?) exercises that are related to commands taught in that lesson. Where possible, students should be given additional problems that can be verified independently by experiment or analytical methods. Students really don't feel comfortable or confident until they can make models from scratch on their own. To that end, in the previous edition a number of new exercise questions were added in Chapters 3 through 10.

That having been said, I am continually amazed at how quickly many students can get up the learning curve on both Creo Parametric and Simulate. Any instructor introducing this software to a group of capable and interested students should be prepared to move very quickly to stay ahead of the class!

Organization and Synopsis of the Tutorials

A brief synopsis of the chapters in this book is given below. Each chapter should take at least 1.5 to 3 hours to complete - if you go through the lessons too quickly or thoughtlessly, you may not understand or remember the material. For best results, it is suggested that you scan/browse through the lesson or major section completely before going through it in detail. You will then have a sense of where the lesson is going, and not be tempted to just follow the commands blindly. You need to have a sense of the forest when examining each individual tree! Another suggestion is to work through the lessons (in Chapters 3 through 10) with a study partner. This helps locate commands on the screen and the inevitable discussion will reinforce learning of the material. The exercises, of course, should be done individually.

Chapter 1 - Introduction to Creo Simulate

This section presents a short introduction to finite element analysis, with some cautions about its use and misuse. Examples of problems solved with Simulate are presented. Some tips and tricks for using the Simulate interface are also covered.

Chapter 2 - Finite Element Modeling with Simulate

This chapter presents an overview of the theory behind how FEA is implemented in Simulate. The primary purpose of this chapter is to outline the main capabilities of Simulate as they apply to the design and analysis of mechanical parts. These include simple analyses,

---

[2] Is this the result or effect of complicated computer games and how they are learned? Is this a variation of "discovery learning" and, if so, is it that bad?

sensitivity studies, and parameter optimization. This chapter will basically introduce you to the terminology used in the program, and give you an overview of its operation.

Chapter 3 - Solid Models (Part 1 - Static Analysis)

By analyzing a very simple solid model, this chapter presents the basic procedure and commands for performing static stress analysis. This is a natural starting point, given that models imported from Creo are usually solids. Common methods of displaying results are shown, including deformation animation, stress fringe plots, convergence graphs, and cutting surfaces. Automatic mesh generation and some simple mesh controls are introduced.

Chapter 4 - Solid Models (Part 2 - Sensitivity Studies and Optimization)

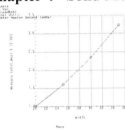

Sensitivity studies and design optimization are important capabilities of Simulate. This chapter explores the creation of parameters and how they are used in these two types of design study. Multiple load sets and superposition are introduced. Special concerns for applying loads and constraints on solid models are explored. Some modeling pitfalls, which also occur in other model types, are investigated and solutions proposed.

Chapter 5 - Plane Stress and Plane Strain

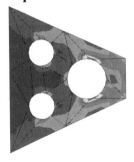

This is the first lesson on idealizations. It deals with problems that can be classed as either plane stress or plane strain. In either case, the model is based on 2D geometry that can be extracted from the Creo solid model. The idealization, when applicable, results in a significant reduction in the computational effort for the model. The use of symmetry is introduced.

Chapter 6 - Axisymmetric Solids and Shells

Axisymmetric models are another case where a 2D idealization can be used. Two are available: axisymmetric solids and shells. New load types are introduced: centrifugal and thermal loads.

Chapter 7 - Shell Models

Shell models are an idealization for general 3D models that can be used when a part is composed of thin-walled features. Shell geometry can be created automatically or manually. Shells can also be combined with solids in the same model. Bearing loads are introduced and more convergence problems are discussed.

Chapter 8 - Beams and Frames

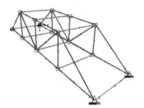

Beam elements can be used to form continuous straight and curved beams, and 2D/3D frames and trusses. The use of beam elements requires a good understanding of beam coordinate systems, sections, and orientation. Point and distributed loads are covered, as are beam releases. The preparation of shear and bending moment diagrams for individual beams is demonstrated. Beams can also be used in combination with shells and solids.

Chapter 9 - Miscellaneous Topics

Several topics are introduced here, starting with cyclic symmetry. The use of springs and masses is examined. Modal analysis is introduced. Finally, the use of bonded and contact interfaces in a simple assembly is demonstrated.

Chapter 10 - Thermal Analysis

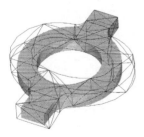

The final chapter will introduce methods in *Thermal* to solve steady state and transient heat transfer problems in 2D and 3D geometries. The importance of units in thermal analysis is emphasized. Thermal analysis performed on solid or 2D models allows the determination of temperature distributions and heat fluxes through the model boundaries. Transient thermal problems must be performed on solid models. Once the thermal results are obtained, the temperature solution can be transferred back to Structure to determine thermally induced stresses.

Acknowledgments

Some of the models used in these lessons are based on the treatment in **The Finite Element Method in Mechanical Design** (PWS-Kent, 1993) by Charles E. Knight. This is a clearly written and informative book, although emphasis is on the h-code analysis with only tangential mention of the p-code method used in Creo. Results obtained here can sometimes be compared with those obtained by Knight.

Several of the exercise questions are variations of problems taken from the book **A First Course in the Finite Element Method, 4th Ed.** (Thomson, 2007) by Daryl Logan.

Thanks are due as always to Stephen Schroff at SDC Publications for his efforts in taking this work to a wider audience. Also to Zach, Karla, and Megan for their patience in waiting for the manuscript to arrive and post-processing of the manuscript.

Once again I must express special thanks to Elaine, Jenny, and Kate for their patience while I was occupied with this work.

To users of this material, I hope you enjoy the lessons. I apologize beforehand for any omissions and errors that may have appeared and I would appreciate any comments, criticisms, and suggestions for the improvement of this manual.

RT
Edmonton, Alberta
6 May 2019

This page left blank.

TABLE OF CONTENTS

Chapter 1 - Introduction to FEA

Chapter 2 - Finite Element Analysis with Creo Simulate

Chapter 3 - Solid Models - Part 1 (Standard Static Analysis)

Chapter 4 - Solid Models - Part 2 (Design Studies, Optimization, AutoGEM Controls, Superposition)

Chapter 5 - Plane Stress and Plane Strain Models

Chapter 6 - Axisymmetric Solids and Shells

Chapter 7 - Shell Models

Chapter 8 - Beams and Frames

Chapter 9 - Miscellaneous Topics (Cyclic Symmetry, Modal Analysis, Springs and Masses, Contact Analysis)

Chapter 10 - Thermal Models (Steady state and transient models; transferring thermal results for stress analysis)

This page left blank.

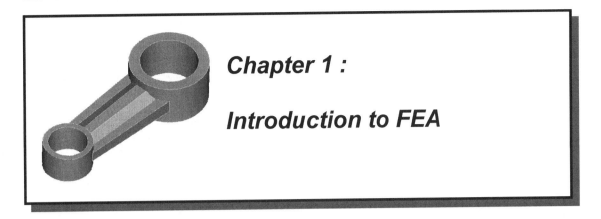

Overview

This lesson will be used to get you set up for the rest of the tutorials. It will go over some basic ideas about FEA and what you can do with Creo Simulate. The chapter is quite short, and will cover the following:

♦ general comments about using Finite Element Analysis (FEA)
♦ examples of problems solved using Creo Simulate
♦ how the tutorial will present command sequences
♦ some tips and tricks for using Creo Simulate

Finite Element Analysis

Finite Element Analysis (FEA), also known as the Finite Element Method (FEM), is arguably the most important addition to the design engineer's toolkit in the last few decades. The development of FEA has been driven by the desire for more accurate design computations in more complex situations, allowing improvements in both the design procedure and products. The use of simulation in design (FEA in particular) was made possible by the creation of affordable computers that are capable of handling the immense volume of calculations necessary to prepare and carry out an analysis and easily display the results for interpretation. FEA is available at a practical cost to virtually all engineers and designers.

The Creo Simulate[1] software described in this introductory tutorial is only one of many commercial FEA systems that are available. All of these systems share many common capabilities. In this tutorial, we will try to present both the commands for using Creo and the reasons behind those commands so that the general ideas might be transferred to other FEA packages. Notwithstanding this desire, it should be realized that Creo is unique in many ways among software currently available. Therefore, numerous topics treated will be specific to Creo Simulate.

---

[1] This will be referred to as "Creo" from now on, unless reference is being made to other software in the Creo family, like Creo Parametric or Creo Direct.

Creo Simulate is able to perform the following *structural* analyses:

- static stress analysis for linear materials
- modal analysis (mode shapes and natural frequencies)
- buckling analysis
- fatigue analysis
- stress analysis of composite materials
- non-linear static stress analysis (hyperelastic materials and/or large deformations)

and others. This tutorial will be concerned only with the first two of these analyses, the remaining types of problems being beyond the scope of an introductory manual. Once having finished these lessons, however, interested users should be able to take on the more complex models with more confidence. Creo can also perform the following *thermal* analyses:

- steady state and transient analysis of problems involving conduction, convection, and radiation.

The results of the thermal analysis (computed temperature distributions) can be used to compute the associated thermal stresses. We will have a brief look at thermal analyses later in these lessons.

A previous incarnation (PTC's MECHANICA) contained a third program, *Motion*, which was used for dynamic analysis of mechanisms. Some of the functionality of that program has been included within Creo Parametric with the appropriate license configuration. Full dynamic simulation capabilities are available which allows passing information (dynamic loads) back to Creo Simulate in order to compute associated stresses. This tutorial will not cover any mechanism kinematics or dynamics.

Creo offers much more than simply an FEA engine. We will see that it is really an easily used but powerful design tool since it will allow parametric studies as well as design optimization to be set up quite easily. Moreover, unlike many other commercial FEM programs where determining accuracy can be difficult or time consuming, Creo Simulate will be able to compute results with some certainty as to the accuracy[2].

As mentioned above, Creo can handle problems involving non-linearly elastic (hyperelastic) materials like rubber. Problems involving very large geometric deflections (that is, geometric non-linearity) can also be treated. Creo can also treat problems where the material is loaded beyond its elastic limit (i.e. elastoplastic deformation) and therefore undergoes permanent, non-recoverable deformation.

---

[2] This refers to the problem of "convergence" whereby the FEA results must be verified or tested so that they can be trusted. We will discuss convergence at some length later on and refer to it continually throughout the manual.

In this tutorial, we will concentrate on the main concepts and procedures for using the software and focus on topics that seem to be most useful for new users and/or students doing design projects and other course work. We assume that readers do not know anything about simulation but are quite comfortable with Creo Parametric (in particular the user interface). A short and very qualitative overview of the FEA theoretical background has been included, and it should be emphasized that this is very limited in scope. Our attention here is on the use and capabilities of the software, not providing a complete course on the theoretical origins of FEA. For further study of these subjects, see the reference list at the end of the second chapter.

Examples of Problems Solved using Creo Simulate

To give you a taste of what is to come, here are four examples of what you will be able to do on completion of this tutorial. The first three examples are structural problems and include a simple analysis, a parametric study called a sensitivity analysis, and a design optimization. The fourth example is a thermal analysis where the temperature distribution is brought back into a stress analysis to compute thermally induced stresses.

Example #1 : Stress Analysis

This is the "bread and butter" type of problem for Creo. A model is defined by some geometry (in 2D or 3D) in the geometry pre-processor (Creo Parametric). This is not as simple or transparent as it sounds, as discussed below. The model is transferred into Creo Simulate where material properties are specified, loads and constraints are applied, and one of several different types of analysis can be run on the model. In the figure at the right, a model of a somewhat crude connecting rod is shown. This part is modeled using 3D solid elements. The surface of the hole at the large end is fixed and a lateral bearing load is applied to the inside surface of the hole at the other end.

Figure 1 Solid model of a part

The primary results are shown in Figures 2 and 3. These are contours of the Von Mises stress[3] on the part, shown in a *fringe* plot (these are, of course, in color on the computer screen), and a wireframe view of the total (exaggerated) deformation of the part (this can be shown as an animation). Here, we are usually interested in the value and location of the maximum Von Mises stress in the part, whether the solution agrees with our desired boundary conditions, and the magnitude and direction of deformation of the part.

---

[3] The Von Mises stress is obtained by combining all the stress components at a point in a way which produces a single scalar value that is compared to the yield strength of the material to determine failure. This is a common way in FEA of examining the computed stress in a part.

Figure 2 Von Mises stress fringe
plot

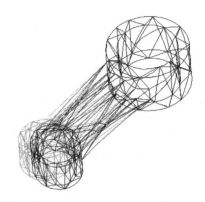

Figure 3 Deformation of the part

Example #2 : Sensitivity Study

Often you need to find out the overall effect on the solution of varying one or more
design parameters, such as dimensions. You could do this by performing a number of
similar analyses, and changing the geometry of the model between each analysis. Creo
has an automated routine which allows you to specify the parameter to be varied, and the
overall range. It then automatically performs all the modifications to the model, and
computes results for the designated intermediate values of the design parameters. This is
possible due to the tight integration with Creo Parametric, which is the geometry engine.

The example shown in Figure 4 is a quarter-model (to take advantage of symmetry) of a
transition between two thin-walled cylinders. The transition is modeled using shell
elements.

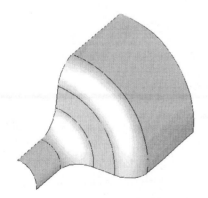

Figure 4 3D Shell quarter-model
of transition between cylinders

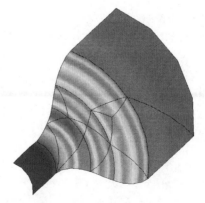

Figure 5 Von Mises stress in
shell model

Figure 5 shows the contours of the Von Mises stress on the part produced by an internal
pressure. The maximum stress occurs at the edge of the fillet on the smaller cylinder just
where it meets the intermediate flat portion. The design parameter to be varied is the
radius of this fillet, between the minimum and maximum shapes shown in Figures 6 and
7.

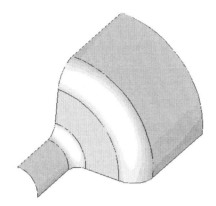

Figure 6 Minimum radius fillet

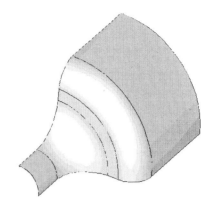

Figure 7 Maximum radius fillet

Figure 8 shows the variation in the maximum Von Mises stress in the model as a function of radius of the fillet. This graph is created automatically for a sensitivity study - a great time saver for the designer. Other information about the model can also be plotted, such as total mass or maximum deflection also as a function of the fillet radius.

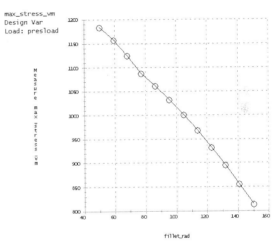

Figure 8 Variation of Von Mises stress with fillet radius in shell model

Example #3 : Design Optimization

This capability of Creo Simulate is really astounding! When a model is created, some of the geometric parameters can be designated as design variables. Then Creo is turned loose to find the combination of values of these design variables that will minimize some objective function (like the total mass of the model) subject to some design constraints (like the allowed maximum stress and/or deflection). Creo searches through the design space (the specified ranges of the design variables) and will find the optimum set of design variables automatically!

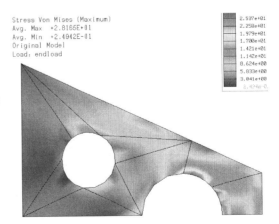

Figure 9 Initial Von Mises stress distribution in plate before optimization

The example shown is of a plane stress model of a thin, symmetrical, tapered plate under tension. The plate is fixed at the left edge, while the lower edge is along the plane of symmetry. A uniform tensile load is applied to the vertical edge on the right end. The Von Mises stress contours for the initial design are shown in Figure 9. The maximum stress, which exceeds a design constraint, has occurred at the large hole on the right, at about the 12:30 position. The stress level around the smaller hole is considerably less, and we could probably increase the diameter of this hole in order to reduce mass. The question is: how much?

The selected design variables are the radii of the two holes. Minimum and maximum values for these variables are indicated in the Figures 10 and 11. The objective of the optimization is to minimize the total mass of the plate, while not exceeding a specified maximum stress.

Figure 10 Minimum values of design variables

Figure 11 Maximum values of design variables

Figure 12 shows a history of the design optimization computations. The figure on the left shows the maximum Von Mises stress in the part that initially exceeds the allowed maximum stress, but Creo Simulate very quickly adjusts the geometry to produce a design within the allowed stress. The figure on the right shows the mass of the part. As the optimization proceeds, this is slowly reduced until a minimum value is obtained (approximately 20% less than the original). Creo Simulate allows you to view the shape change occurring at each iteration.

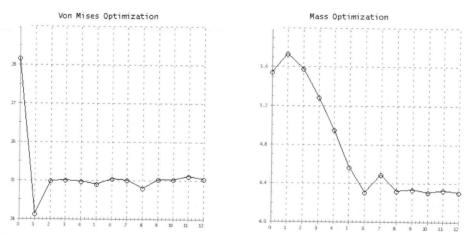

Figure 12 Optimization history: Von Mises stress (left) and total mass (right)

The final optimized design is shown in Figure 13. Notice the increased size of the interior hole, and the more efficient use of material. The design limit stress now occurs on both holes.

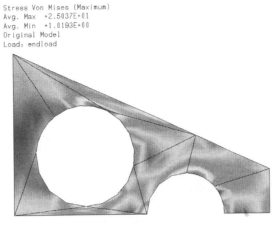

Figure 13 Von Mises stress distribution in optimized plate

In these lessons, we will explore variations of these three types of design study (simple analysis, sensitivity studies, and optimization). We will explore a number of different types of models while doing this (solids, shells, beams, plates, etc.).

Example #4 : Thermally Induced Stress

This example concerns the analysis of the body of a gate valve, shown in Figure 14. The model is first simplified using symmetry and removal of non-essential features (flange bolt holes, o-ring grooves, and some rounds). The resulting model is loaded as follows:

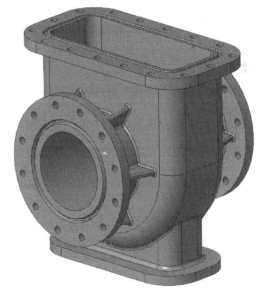

- an internal pressure that acts directly on the interior walls and produces a net upward force on the top flange (caused by the valve bonnet)
- temperature variation in the valve body due to the convective heat transfer caused by a hot gas flowing through the valve, plus heat lost to the environment.

Figure 14 Valve body to be loaded with internal pressure and thermally induced stress

Figure 15 shows the temperature distribution on the model obtained with the thermal analysis. The thermal analysis uses the same mesh as the stress analysis, so that the temperature data can be fed back into the stress analysis and the stresses due to the temperature variation in the model can be determined.

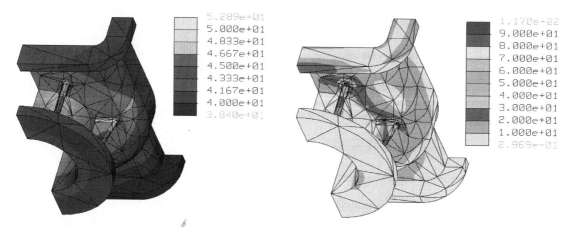

Figure 15 Temperature distribution in the valve body (°C)

Figure 16 Von Mises stress resulting from combined pressure and thermal load

Figure 16 shows the final combined results of the thermal stresses and those due to pressure. Using superposition principles, it is possible to examine these separately to find out which one contributes most to the overall stress (it is the thermal stress in this problem).

This is a static analysis only. More complicated transient heat transfer models can be treated. Thermal model idealizations are available for 2D problems (analogous to plane stress and plane strain models), and beams and shells can also be treated. Some special care must be taken when dealing with convective heat transfer from these idealized geometries. Although Creo can handle radiation heat transfer problems, that type of problem will not be dealt with here.

FEA User Beware!

Users of this (or any other FEA) software should be cautioned that, as in other areas of computer applications, the GIGO ("Garbage In = Garbage Out") principle applies. Users can easily be misled into blind acceptance of the answers produced by the programs. **Do not confuse pretty graphs and pictures with correct modeling practice and accurate results**.

A skilled practitioner of FEA must have a considerable amount of knowledge and experience. The current state of sophistication of CAD and FEA software may lead non-wary users to dangerous and/or disastrous conclusions. Users might take note of the fine print that accompanies all FEA software licenses[4], which usually contains some text along these lines: "The supplier of the software will take no responsibility for the results

---

[4] Or, for example, see the Disclaimer at the front of this book.

obtained . . ." and so on. Clearly, the onus is on the user to bear the burden of responsibility for any conclusions that might be reached from the FEA.

We might plot the situation something like Figure 17. In order to intelligently (and safely) use FEA, it is necessary to acquire some knowledge of the theory behind the method, some facility with the available software, and a great deal of modeling experience. In this manual, we assume that the reader's level of knowledge and experience with FEA initially places them at the origin of the three axes in the figure. The tutorial (particularly Chapter 2) will extend your knowledge a little bit in the "theory" direction, at least so that we can know what the software requires for input data, and (generally) how it computes the results. The step-by-step tutorials and exercises will extend your knowledge somewhat farther in the "experience" direction. Primarily, however, this tutorial is meant to extend your knowledge in the "Creo software" direction. Readers who have already moved out along the "theory" or "experience" axes will have to bear with us (we will not cover a lot of theory, and the problems we will look at are very simple) - at least these lessons should help you discover the capabilities of the Creo Simulate software package.

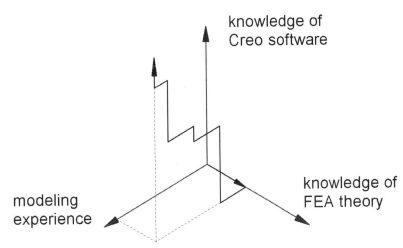

Figure 17 Requirements for Creo users and the Tutorial path

Some quotes from speakers at an FEA panel at an ASME Computers in Engineering conference in the early 1990s should be kept in mind:

"Don't confuse convenience with intelligence."
> In other words, as more powerful functions get built in to FEA packages (such as automatic or even adaptive mesh generation), do not assume that these will be suitable for every modeling situation, or that they will always produce trustworthy results. If an option has defaults, be aware of what they are and their significance to the model and the results obtained. Above all, remember that just because it is easy, it is not necessarily right!

"Don't confuse speed with accuracy."
> Computers are getting faster and faster and it is tempting to infer that

improved technology yields improved results. This also means that they can compute an inaccurate model faster than before - a wrong answer in half the time is hardly an improvement!

and finally, the most important:

"FEA makes a good engineer better and a poor engineer dangerous."

As our engineering tools get more sophisticated, there is a tendency to rely on them more and more, sometimes to dangerous extremes. Relying solely on FEA for design verification is dangerous. Don't forget your intuition, and remember that a lot of very significant engineering design work has occurred over the years on the back of an envelope. Let FEA extend your design capability, not define it.

Tips for using Creo Simulate

In the tutorial examples that follow, you will be lead through a number of simple problems keystroke by keystroke. Each command will be explained in depth so that you will know the "why" as well as the "what" and "how". Resist the temptation to just follow the keystrokes - you must think hard about what is going on in order to learn it. You should go through the tutorials while working on a computer so that you experience (and think about!) the results of each command as it is entered. Not much information will sink in if you just read the material. We have tried to capture exactly the key-stroke, menu selection, or mouse click sequences to perform each analysis. These actions are indicated in *bold face italic type*.

Creo incorporates an "object-action" operating paradigm. This means you can pick an object on the screen (like a part surface), then specify the action to be performed on it (like applying a load). This is a much more streamlined and natural sequence to process commands. Of course, the previous "action-object" form will still work. In this Tutorial, command sequences are represented at various times in either of the two forms. Hopefully, this will not get confusing.

The interface for Creo is based on a ribbon style. As mentioned above, it is assumed that users are familiar with the Creo ribbon interface style. Generally speaking, the ribbons are laid out to follow the work flow. At any stage in model development you will likely find all commands on the same ribbon. In the following, commands will be identified by group and command name (and sometimes icon), with new ribbons identified as required. Any menu commands grayed out are unavailable for the current context. Otherwise, any menu item is available for use.

As of Creo Simulate 4.0 and as part of the object/action command structure, whenever something is selected in the graphics window (geometric entity like surface, edge, or curve, or a modeling entity like a constraint or load icon) or model tree, a mini toolbar menu opens at the cursor location. This is context sensitive and complete enough that you may not need to use the ribbon at all.

 Many operations can be launched by clicking and holding down the right mouse button (RMB) on an entry in the model tree or in the graphics window. This will produce a (context sensitive) pop-up menu of relevant commands.

Characters entered from the keyboard are enclosed within square brackets. When more than one command is given in a sequence, they are separated by the symbol " > ". Selections or settings made within a single menu are separated with " | ".

At the end of each chapter in the manual, we have included some Questions for Review and some simple Exercises which you should do. These have been designed to illustrate additional capabilities of the software, some simple modeling concepts, and sometimes allow a comparison with either analytical solutions or with alternative modeling methods. The more of these exercises you do, the more confident you can be in setting up and solving your own problems.

Finally, for those not familiar with the Creo interface, here are a few hints about using the mouse. Menu items and/or graphics entities on the screen are selected by clicking on them with the *left mouse button*. We will often refer to this as a 'left click' or simply as a 'click'. The *middle mouse button* ('middle click') can be used (generally) whenever ***Accept, Enter, Close*** or ***Done*** is required. The dynamic view controls are obtained using the mouse as shown in Table 1. Users of Creo Parametric will be quite comfortable with these mouse controls.

Table 1-1 Common Mouse Functions

Function		Operation	Action
Selection (click left button)		LMB	entity or command under cursor selected
Direct View Control (drag holding middle button down)		MMB	Spin
		Shift + MMB	Pan
		Ctrl + MMB (drag vertical)	Zoom
		Ctrl + MMB (drag horizontal)	Rotate around axis perpendicular to screen
		Roll MMB scroll wheel (if available)	Zoom
Pop-up Menus (click right button)		RMB with item selected	launch context-sensitive pop-up menus

So, with all that out of the way, let's get started. The next chapter will give you an overview of FEA theory, and how Creo Simulate is different from other commercial packages.

Questions for Review

1. In Creo Simulate language, what is meant by a "design study?"
2. What are the three types of design study that can be performed by Creo Simulate?
3. What is the Von Mises stress? From a strength of materials textbook, find out how this is computed and its relation to yield strength. Also, for what types of materials is this a useful computation?
4. Can Creo Simulate treat non-linear problems?
5. What does GIGO mean?
6. What three areas of expertise are required to be a skilled FEA practitioner?

Exercises

1. Find some examples of cases where seemingly minor and insignificant computer-related errors have resulted in disastrous consequences.
2. Find out what role computers played in the Apollo missions that landed the first astronauts on the moon on July 20, 1969. (Hint: a modern smart phone probably contains more computing power than was available to the design engineers of the day!)
3. a) Simple analytical models result in the following formulas for simple states of stress:

$$\sigma = \frac{P}{A} \qquad \sigma = \frac{My}{I} \qquad \tau = \frac{Tc}{J} \qquad \tau = \frac{VQ}{It}$$

   Identify each of these formulas and list the assumptions that were used in their derivation. For each formula, **which of these assumptions is most likely to be violated in a real-world problem**? Which assumptions do you think are also used in the finite element analysis of a 3D solid model?
   b) What is the formula for Von Mises stress? Where does this come from, and why are we interested in it?
4. What is meant by the *principle of superposition*, and what does it have to do with stress analysis? (We will need this in Chapter 4.)
5. Find out something about the history of the development of FEA. When was it "invented" and what kinds of problems was it originally used for?

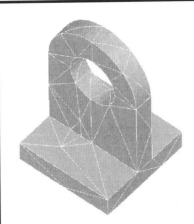

Chapter 2 :

Finite Element Analysis with Creo Simulate

Overview of this Lesson

This chapter presents an overall view of FEA in general, and discusses a number of ideas and issues involved. The major differences between Creo, which uses a p-code method, and other packages, which typically use h-code, are presented. The topics of accuracy and convergence are discussed. The major sections in this chapter are:

- ◆   overview and origins of FEA
- ◆   discussion of the concept of the "model"
- ◆   general procedure for FEA solutions
- ◆   FEA models versus CAD models
- ◆   p-elements and h-elements
- ◆   convergence and accuracy
- ◆   sources of error
- ◆   overview of Creo

Although you are probably anxious to get started with the software, your understanding of the material presented here is very important. We will get to the program soon enough!

Finite Element Analysis : An Introduction

In this section, we will try to present the essence of FEA without going into a lot of mathematical detail. This is primarily to set up the discussion of the important issues of accuracy and convergence later in the chapter. Some of the statements made here are generalizations and over-simplifications, but we hope that this will not be too misleading. Interested users can consult any of a number of text and reference books (some are listed at the end of this chapter) which describe the theoretical underpinnings of FEA in considerably greater and rigorous detail.

In the following, the ideas are illustrated using a planar (2D) solution region, but of course these ideas extend also to 3D. We'll try to keep the mathematics to a minimum here! Let's suppose that we are faced with the following problem: We are given a body or connected region (or volume) R with a boundary B as shown in Figure 1(a). Both interior and exterior boundaries might be present and can be arbitrarily shaped. Some continuous physical variable, for example temperature T or displacement u, is governed by a physical law within the body and subjected to known conditions on the boundary. This physical law might be expressed using the theory of heat conduction or elasticity. We wish to find the values of T (or u) within the region.

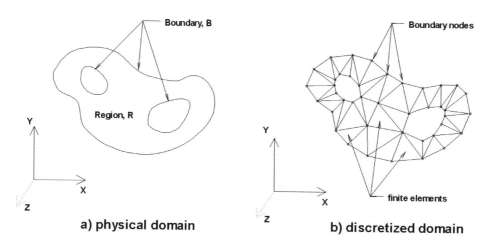

a) physical domain   b) discretized domain

Figure 1 The problem to be solved is specified in a) the physical domain and b) the discretized domain used by FEA

For a two dimensional problem, the governing physical law or principle might be expressed by a partial differential equation (PDE), for example[1]:

$$\frac{\partial^2 T}{\partial x^2} + \frac{\partial^2 T}{\partial y^2} = 0$$

that is valid in the interior of the region R. Note that this governing PDE is usually the result of simplifying assumptions made about the physical system, such as the material being homogeneous and isotropic, with constant linear properties, and so on. The solution to the PDE must satisfy some boundary conditions or *constraints* prescribed on the boundary B. These might be in the form of specified values or specified gradients normal to the boundary.

In a finite element solution, the geometry of the region is typically generated by a CAD program, such as Creo Parametric.

In order to solve this problem, the region R is *discretized* (divided) into individual *finite elements* that collectively approximate the shape of the region, as shown in Figure 1(b). This discretization is accomplished by locating *nodes* along the boundary and in the

---

[1] The PDE given represents the steady state temperature distribution within a planar body which is governed by conduction. There are no heat sources within the body.

interior of the region. The nodes are then joined by lines to create the finite elements. In 2D problems, these can be triangles or quadrilaterals; in 3D problems, the elements can be tetrahedra or 8-node "bricks". In some FEA software, other higher order types of elements are also possible (e.g. hexagonal prisms). Some higher order elements also have additional nodes along their edges. Collectively, the set of all the elements is called a *finite element mesh*. In the early days of FEA, a great deal of manual effort was required to set up the mesh. More recently, automatic meshing routines have been developed in order to do most, if not all, of this tedious task.

In the FEA solution, values of the dependent variable (T, in our example) are computed only at the nodes. The variation of the variable within each element is determined from the nodal values so as to **approximately** satisfy the governing PDE[2]. One way of doing this is by using interpolating polynomials. In order for the PDE to be satisfied, the nodal values of each element must satisfy a set of conditions represented by several linear algebraic equations involving other nodal values.

In the simplest case, the boundary conditions are implemented by specifying the values of the variables on the boundary nodes. There is no guarantee that the true boundary conditions on the continuous boundary B are satisfied between the nodes on the discretized boundary.

When all the individual elements in the mesh are combined, the discretization and interpolation procedures result in a conversion of the problem from the solution of a continuous differential equation into a very large set of simultaneous linear algebraic equations. This system will typically have many (sometimes hundreds of) thousands of equations in it. The solution of this algebraic system requires special and efficient numerical algorithms. It is important to remember that the solution contains the nodal values that collectively represent an *approximation* to the continuous solution of the initial PDE. An important issue, then, is the accuracy of this approximation. In classical FEA solutions, the approximation becomes more accurate as the mesh is refined with smaller elements. In the limit of zero mesh size, requiring an infinite number of equations, the FEA solution to the PDE would be exact. This is, of course, not achievable. So, a major issue revolves around the question "How fine a mesh is required to produce answers of acceptable accuracy?" and the practical question is "Is it feasible to compute this solution?" We will see a bit later how Creo solves these problems.

Once the large set of simultaneous linear equations is solved, the results of interest can be examined and displayed. Sometimes this is the main solution variable itself (T in our example above); sometimes it is a quantity derived from the solution variable (like heat flux, which depends on the gradient of the temperature field). This step, known as post-processing, is where the graphics display power of computers is most evident. It is also where post-processing magic can lead to erroneous conclusions!

---

[2] Someone has suggested that a more accurate definition of FEA is "Finite Element Approximation".

IMPORTANT POINT: In FEA stress analysis problems, the dependent variable in the governing PDEs is the *displacement* from the reference (usually unloaded) position. The material strain (displacement per unit length) is then computed from the displacement by taking the derivative with respect to position. Finally, the stress components at any point in the material are computed from the strain at that point. Thus, if the interpolating polynomial for the spatial variation of the displacement field within an element is linear, then the strain and stress will be constant within that element, since the derivative of a linear function is a constant. The significance of this will be illustrated a bit later in this lesson.

The FEA Model and General Processing Steps

Throughout this manual, we will be using the term "model" extensively. We need to have a clear idea of what we mean by the FEA model.

To get from the "real world" physical problem to the approximate FEA solution, we must go through a number of simplifying steps (Figure 2). At each step, it is necessary to make decisions about what assumptions or simplifications will be required in order to reach a final workable model. By "workable", we mean that the FEA model must allow us to compute the results of interest (for example, the maximum stress in the material) with sufficient accuracy and with available time and resources. It is no good building a model that is over-simplified to the point where it cannot produce the results with sufficient accuracy. It is also no good producing a model that is "perfect" but requires a

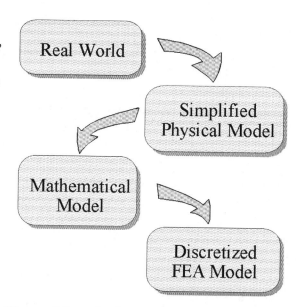

Figure 2 Developing a Model in Finite Element Analysis

supercomputer and will not yield useful computational results for several weeks! Quite often, the FEA user must compromise between the two extremes - accepting a slightly less accurate answer in a reasonable solution time. The trick is to know exactly (or even approximately) how much accuracy you are trading off for the reduced processing time. Fortunately, Creo makes this determination a part of the solution itself.

Real World → Simplified Physical Model

This simplification step involves making assumptions about physical properties or the physical layout and geometry of the problem. For example, we usually assume that materials are homogeneous and isotropic and free of internal defects or flaws. It is also common to ignore aspects of the geometry that will have no (anticipated) effect on the results, such as the chamfered and filleted edges on the bracket shown in Figure 3, and perhaps even the mounting holes themselves. Ignoring these "cosmetic" features, as shown in Figure 4, is often necessary in order to reduce the geometric complexity so that the resulting FEA model is practical. On the other hand, care must be taken that seemingly minor, but ultimately essential features of the geometry are not inadvertently eliminated (for example, the fillet where the two plates meet).

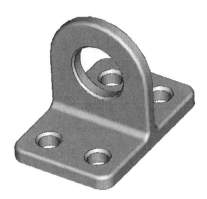

Figure 3 The "Real World" Object

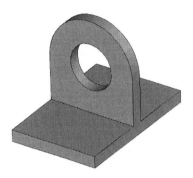

Figure 4 The idealized physical model

Simple Physical Model → Mathematical Model

To arrive at the mathematical model, we make assumptions like linearity of material properties, idealization of loading conditions, and so on, in order to apply mathematical formulas to complex problems. We often assume that loading is steady, that fixed points are perfectly fixed, beams are long and slender, and so on. As discussed above, the mathematical model usually consists of one or more differential equations that describe the variation of the variable of interest within the boundaries of the model. For very simple geometric models, there may be analytical solutions for stress values (see Exercise #3).

Mathematical Model → FEA Model

The simplified geometry of the model is discretized (see Figure 5) so that the governing differential equations obtained in the previous step can be rewritten as a (large) number of simultaneous linear equations representing the assembly of elements in the model.

In the operation of FEA software, the three modeling steps described above often appear to be merged. In fact, most of it occurs below the surface (you will never see the governing PDE, for example) or is inherent in the software itself. For example, Creo automatically assumes that materials are

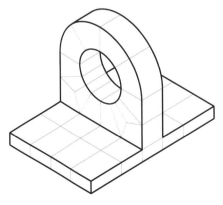

Figure 5 A mesh of solid brick elements

homogeneous, isotropic, and linear. (All of these default assumptions can be overridden, of course.) It is useful to remind yourself about these separate aspects of modeling from time to time, because each is a potential source of error or inaccuracy in the results.

Steps in Preparing an FEA Model for Solution

Starting from the simplified geometric model, there are generally several steps to be followed in the analysis:

1.   identify the model type
2.   specify the material properties, model constraints, and applied loads
3.   discretize the geometry to produce a finite element mesh
4.   solve the system of linear equations
5.   compute items of interest from the solution variables
6.   display and critically review results and, if necessary, repeat the analysis

The overall procedure is illustrated in Figure 6. Some additional detail on each of these steps is given below. The major steps must be executed in order, and each must be done correctly before proceeding to the next step. When a problem is to be re-analyzed (for example, if a stress analysis is to be performed for the same geometry but different loads), it will not usually be necessary to return all the way to the beginning. The available re-entry points will become clear as you move through these tutorials.

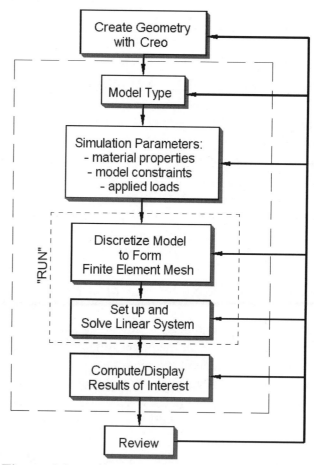

The steps shown in the figure are:

1. The geometric model of the part/system is created using Creo Parametric.

2. On entry to Creo Simulate (functions inside the outer dashed rectangle), the model type must be identified. The default is a solid model.

Figure 6 Overall steps in FEA Solution

3. A) Specify material properties for the model. It is not necessary that all the elements have the same properties. In an assembly, for example, different parts can be

made of different materials. For stress analysis the required properties are Young's modulus and Poisson's ratio. We can also define solids with hyperelastic properties, like rubber. For thermal problems we must include density, thermal conductivity and heat capacity. Most FEA packages include built-in libraries containing properties of common materials (steel, iron, aluminum, etc.).

B) Identify the constraints on the solution. In stress analysis, these could be fixed points, points of specified displacement, or points free to move in specified directions only. In thermal analysis, constraints involve prescribed temperatures or convective heat transfer conditions.

C) Specify the applied loads on the model (point loads, uniform edge loads, pressure on surfaces, direct heat loads, etc.).

4. Once you are satisfied with your model, you set up and run a processor that actually performs the solution to the posed FEA problem. This starts with the automatic creation of the finite element mesh from the geometric model by a subprogram within Creo called AutoGEM. Creo will trap some modeling errors here, such as elements with no material properties assigned to them. The FEA engine then sets up and solves the system of linear equations. Since this system can be very large (even hundreds of thousands of equations), special numerical techniques are used to exploit certain properties of the system. The processor will produce a summary file of output messages which can be consulted if something goes wrong - for example, a model that is not sufficiently constrained by boundary conditions.

5. FEA produces immense volumes of output data. The only feasible way of examining this is graphically. Creo has very powerful graphics capabilities to examine the results of the FEA - displaced shape and animations, stress distributions, mode shapes, etc. Hard copy of the results file and screen display is easy to obtain. You can even directly produce web pages with result images and comments, and animations of solutions (deformation or heat flow). It is possible, and common practice, to "instrument" the model by requesting *measures* of specific quantities (stress, displacement, temperature, . . .) at specific locations in the model or the maximum/minimum values occurring in a designated region.

6. Finally, the results must be reviewed critically. In the first instance, the results should agree with our modeling intent. For example, if we look at an animated view of the deformation, we can easily see if our boundary constraints have been implemented properly. The results should also satisfy our intuition about the solution (stress concentration around a hole, for example). If there is any cause for concern, it may be advisable to revisit some aspects of the model and perform the analysis again (with slightly different parameters) to explore the model's behavior. This can often identify flaws in how the model was set up[3].

---

[3] You should try to learn something from every run of the model. As stated by Richard Hamming in his book <u>Numerical Methods for Scientists and Engineers</u> (1962): "The purpose of computing is insight, not numbers."

P-Elements versus H-Elements

Not all discretized finite elements are created equal! Here is where a major difference arises between Creo and most other FEA programs.

Convergence of H-elements (the "classic" approach)

Following the classic approach, other programs often use low order interpolating polynomials in each element. This has significant ramifications, especially in stress analysis. As mentioned above, in stress analysis the primary solution variables are the displacements of the nodes. The interpolating functions are typically linear (first order) within each element. Strain is obtained by taking the derivatives of the displacement field and the stress is computed from the material strain. For a first order interpolating polynomial within the element, this means that the strain and therefore the stress components within the element are constant everywhere. The situation is depicted in Figure 7, which shows the computed Von Mises stress in each of the elements surrounding a hole in a thin plate under tension. Such discontinuity in the stress field between elements is, of course, unrealistic and will lead to inaccurate values for the maximum stress. Low order elements lead to the greatest inaccuracy precisely in the regions of greatest interest, typically where there are large gradients within the real object.

An even more disastrous situation is shown in Figure 8. This is a solid cantilever beam modeled using solid brick elements with a uniform transverse load. With only a single first-order element through the thickness, the computed stress will be the same on the top and bottom of the beam. This is clearly wrong, yet the early FEA literature and product demonstrations abounded with examples similar to this.

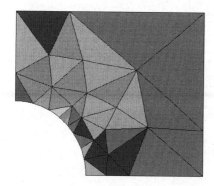

Figure 7 Von Mises stress in 1/4 model of thin plate under tension using first order elements

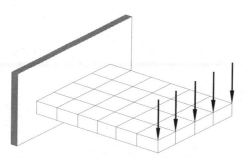

Figure 8 A disaster waiting to happen using first order elements

This situation is often masked by the post-processing capabilities of the software being used, which will sometimes average or interpolate contour values within the mesh or perform other "smoothing" functions strictly for visual appearance. This is strictly a post-processing step, and may bear no resemblance at all to what is actually going on in the mathematical model or (especially) in the real object.

When using first order elements, then, in order to get a more accurate estimate of the stress, it is necessary to use much smaller elements, a process called *mesh refinement*. It may not always be possible to easily identify regions where mesh refinement is required, and quite often the entire mesh is modified. The process of mesh refinement continues until further mesh division and refinement does not lead to significant changes in the obtained solution. The process of continued mesh refinement leading to a "good" solution is called *convergence analysis*. Of course, in the process of mesh refinement, the size of the computational problem becomes larger and larger and we may reach a limit for practical problems (due to time and/or memory limits) before we have successfully converged to an acceptable solution.

The use of mesh refinement for convergence analysis is inherent in the *h-element* class of FEA methods. This "h" is borrowed from the field of numerical analysis, where it denotes the fact that convergence and accuracy are related (sometimes proportional to) the step size used in the solution, usually denoted by *h*. In FEA, the *h* refers to the size of the elements. The elements, always of low order, are referred to as h-elements, and the mesh refinement procedure is called *h-convergence*. This situation is depicted in parts (a) and (b) of Figure 9, where a series of constant-height steps is used to approximate a smooth continuous function. The narrower the steps, the more closely we can approximate the smooth function. Note also that where the gradient of the function is large (such as near the left edge of the figure), then mesh refinement will always produce increasingly higher maximum values.

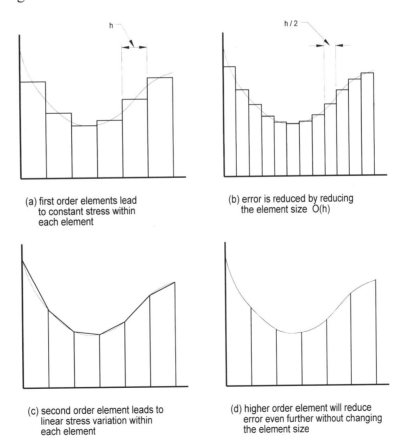

(a) first order elements lead to constant stress within each element

(b) error is reduced by reducing the element size O(h)

(c) second order element leads to linear stress variation within each element

(d) higher order element will reduce error even further without changing the element size

Figure 9 Approximation of stress function in a model

The major outcome of using h-elements is the need for meshes of relatively small elements. Furthermore, h-elements are not very tolerant of shape extremes in terms of skewness, rapid size variation through the mesh, large aspect ratio, and so on. This further increases the number of elements required for an acceptable mesh, and this, of course, greatly increases the computational cost of the solution.

Convergence of P-elements (the Creo approach)

The major difference incorporated in Creo is the following: instead of constantly refining and recreating finer and finer meshes, convergence is obtained by *increasing the order of the interpolating polynomials on each element*. The mesh stays the same for every iteration, called a *p-loop pass*. The use of higher order interpolating polynomials for convergence analysis leads to the *p-element* class of FEA methods, where the "p" denotes polynomial. This method is depicted in parts (c) and (d) of Figure 9. Only elements in regions of high gradients are bumped up to higher order polynomials. Furthermore, by examining the effects of going to higher order polynomials, Creo can monitor the expected error in the solution, and automatically increase the polynomial order only on those elements where it is required. Thus, the convergence analysis is performed quite automatically, with the solution proceeding until an accuracy limit (set by the user) has been satisfied. With Creo, the limit for the polynomial order is 9. In theory, it would be possible to go to higher orders than this, but the computational cost starts to rise too quickly. If the solution cannot converge even with these 9th order polynomials, it may be necessary to recreate the mesh in critical regions at a slightly higher density so that lower order polynomials will be sufficient. This is a very rare occurrence.

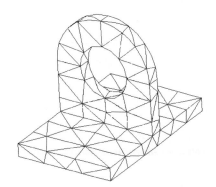

Figure 10 A mesh of solid tetrahedral (4 node) h-elements

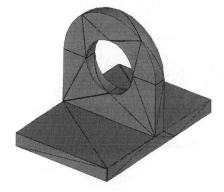

Figure 11 A mesh of tetrahedral p-elements produced by Creo.

The use of p-elements has a number of features/advantages:

▸ The same mesh can be used throughout the convergence analysis, rather than recreating meshes or local mesh refinement required by h-codes. For complicated geometries, eliminating the need for re-meshing results in significantly less computer time.

▸ The mesh is virtually always more coarse and contains fewer elements than h-codes. Compare the meshes in Figures 10 and 11, and note that the mesh of h-elements in

Figure 10 would probably not produce very good results, depending on the loads and constraints applied. The reduced number of elements in Creo (which can be a couple of orders of magnitude smaller) initially reduces the computational load, but as the order of the polynomials gets higher, this advantage is somewhat diminished.

▸ The restrictions on element size and shape are not nearly as stringent for p-elements as they are for h-elements (where concerns of aspect ratio, skewness, and so on often arise).

▸ Automatic mesh generators, which sometimes produce very poor meshes for h-elements, are much more effective with p-elements, due to the reduced requirements and limitations on mesh geometry.

▸ Since the same mesh is used throughout the analysis, this mesh can be tied directly to the geometry. This is the key reason why Creo is able to perform sensitivity and optimization studies during which the geometric parameters of a body can change, but the program does not need to be constantly re-meshing the part.

Convergence and Accuracy in the Solution

It should be apparent that, due to the number of simplifying assumptions necessary to obtain results with FEA, we should be quite cautious about the results obtained. No FEA solution should be accepted unless the convergence properties have been examined.

For h-elements, this generally means doing the problem several times with successively smaller elements and monitoring the change in the solutions. If decreasing the element size results in a negligible (or acceptably small) change in the solution, then we are generally satisfied that the FEA has wrung all the information out of the model that it can.

As mentioned above, with p-elements, the convergence analysis is built in to the program. Since the geometry of the mesh does not change, no expensive (time-wise) remeshing is required. Rather, each successive solution (called a *p-loop pass*) is performed with increasing orders of polynomials (only on elements where this

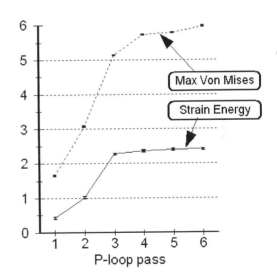

Figure 12 Two common convergence measures using p-elements.

is required) until the change between iterations is "small enough". Figure 12 shows the convergence behavior of two common measures used to monitor convergence in Creo. These are the maximum Von Mises stress and the total strain energy. Note that the Von Mises stress will generally always increase during the convergence test, but can sometimes behave quite erratically as we will see later. Because Von Mises stress is a local measure, the total strain energy is probably a better measure to use to monitor convergence.

Sources of Error

Error enters into the FEA process in a number of ways:

♦ **errors in problem definition** - are the geometry, loads, and constraints known and implemented accurately? Is the correct analysis being performed? Are the material properties correct and/or appropriate?

♦ **errors in creating the physical model** - can we really use symmetry? Is the material isotropic and homogeneous, as assumed? Are the physical constants known? Does the material behave linearly?

♦ **errors in creating the mathematical model** - is the model complete enough to capture the effects we wish to observe? Is the model overly complex? Does the mathematical model correctly express the physics of the problem?

♦ **errors in discretization** - is the mesh too coarse or too fine? Have we left accidental "holes" in the model? If using shell elements, are there tears or rips (free edges) between elements where there shouldn't be?

♦ **errors in the numerical solution** - when dealing with very large computational problems, we must always be concerned about the effects of accumulated round-off error. Can this error be estimated? How trustworthy is the answer going to be?

♦ **errors in interpretation of the results** - are we looking at the results in the right way to see what we want and need to see? Are the limitations of the program understood[4]? Has the possible misuse of a purely graphical or display tool obscured, hidden, or misrepresented a critical result?

You will be able to answer most of these questions by the time you complete this tutorial. The answers to others will be problem dependent and will require some experience and further exposure before you are a confident and competent FEA user.

A CAD Model is *NOT* an FEA Model!

One of the common misconceptions within the engineering community is the equivalence of a CAD solid model with a model used for FEA. These are *not* the same despite proclamations of the CAD vendors that their solid models can be "seamlessly" ported to one or another FEA program. In fact, this is probably quite undesirable! It should not be surprising that CAD and FEA models are different, since the two models are developed for different purposes.

---

[4] The author once had a student who was rightly concerned about the very large deflections in a truss computed using a simple FEA program. The program was performing a linear analysis, and was computing stresses in some members two orders of magnitude higher than the yield strength of the material. The student did not realize that the software knew nothing about failure of the material. It turned out that a simple data entry error had reduced the cross sectional area of the members in the truss.

The CAD model is usually developed to provide a data base for manufacturing. Thus, dimensions must be fully specified (including tolerances), all minor features (such as fillets, rounds, holes) must be included, processing steps and surface finishes are indicated, threads are specified, and so on. Figure 13 shows a CAD solid model of a hypothetical piping component, complete with bolt holes, flanges, o-ring grooves, chamfered edges, and carrying lugs. Not visible in the figure are the dimensions, tolerances, and welding instructions for fabrication which are all part of the CAD model.

Figure 13 A hypothetical 3D solid model of a piping junction

Figure 14 The 3D solid model of pipe junction

FEA is usually directed at finding out other information about a proposed design. To do this *efficiently*, the FEA model can (and often needs to) be quite different from the CAD model. A simple example of this is that the symmetry of an object is often exploited in the preparation of the FEA model. In one of the exercises we will do later, we will model a thin tapered plate with a couple of large holes. The plate has a plane of symmetry so that we only need to do FEA of one-half of the plate. It is also quite common in FEA to ignore minor features like rounds, fillets, chamfers, holes, minor changes in surface profile, and other cosmetic features unless it is expected that these features will have a significant effect on the measures of interest in the model. Most frequently, they do not, and can be ignored.

Figure 14 shows an FEA model of the piping component created to determine the maximum stress conditions in the vicinity of the filleted connection between the two pipes. The differences between the two models shown in Figures 13 and 14 are immediately obvious. Figures 15 and 16 show the mesh of shell elements created from the surface, and the computed Von Mises stress.

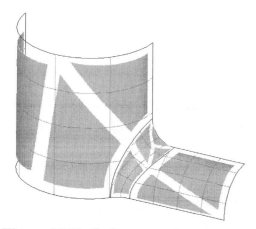

Figure 15 Shell elements of specified thickness created from 3D model

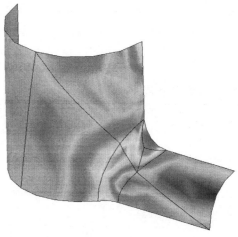

Figure 16 Von Mises stress in the FEA model

In summary, the stated goal of FEA (the "Golden Rule", if you like) might be expressed as:

Use the simplest model possible that will yield sufficiently reliable results of interest at the lowest computational cost.

You can easily see how this might be at odds with the requirements of a CAD model. For further discussion of this, see the excellent book ***Building Better Products with Finite Element Analysis*** by Vince Adams and Abraham Askenazi, Onword Press, 1998.

Overview of Creo Simulate

Basic Operation

We are going to start using Creo in the next chapter. Before we dive in, it will be useful to have an overall look at the function and organization of the software. This will help to explain some of the Creo terminology and see how the program relates to the ideas presented in this chapter's overview of FEA.

We can divide the operation and functionality of Creo according to the rows in Table 2-1 below. These entries are further elaborated in the next few pages. In the process of setting up and running a solution, you will basically need to pick one option from each row in the table. The top-down organization of the table is roughly in the order that these decisions must be made. Other issues such as creation of the model geometry and post-processing and display of final results will be left to subsequent chapters.

Modes of Operation

Creo Simulate can operate in two modes: **integrated** and **standalone**. The user interface is the same for both, so only has to be learned once. Integrated mode is probably the most common way to use Creo Simulate, since the model geometry is created in Creo Parametric and with a click of a button you are switched over to Creo Simulate. When you exit Simulate you are back in Parametric. This tight integration makes it very easy to perform design modifications and quick FEA. Standalone mode allows you to load models directly into Simulate, that is without first loading Parametric. This gives you access to the Simulate FEA modeling tools but not the Parametric geometric modeling tools (e.g. to create or even edit geometric features). When you exit standalone Simulate you close the Creo window. As mentioned above, all the tutorials in this manual are meant to be run in integrated mode.

TABLE 2-1 - An Overall View of Creo Capability and Function

	Creo Options	Description
Mode of Operation	Integrated Standalone	how Creo Simulate is operated with respect to Creo Parametric
Type of Model	3D Plane Stress Plane Strain Axisymmetric	basic structure of the model
Type of Elements	Shell Beam Solid Spring Mass	element types that can be used in a model
Analysis Methods	**Structure:** Static Modal Buckling Pre-stress modal Pre-stress buckling **Thermal:** Steady state Transient	the fundamental solution being sought for the model
Convergence Methods	Quick Check Single Pass Adaptive Multi-Pass Adaptive	method of monitoring convergence in the solution
Design Studies	Standard Sensitivity Optimization	high level methods to organize essentially repetitive computations

Types of Models

This is fairly self-explanatory. In addition to 3D solid, shell, and beam models, Creo in both modes can treat 2D models (plane stress, plane strain, or axisymmetric). Note that all geometry and model entities (loads and constraints) for all 2D model types must be defined in the XY plane of a selected coordinate system. Also, a very thin plate might be modeled as a 2D shell, but if it is loaded with any force components normal to the plate, then it becomes a 3D problem.

Types of Elements

The various types of elements that can be used in Creo are listed in Table 2-1. It is possible to use different types of elements in the same model (e.g. combining solid + beam + spring elements), but we will discuss only a couple of models of this degree of complexity in these tutorials. At first glance, this seems like a limited list of element types. H-element programs typically have large libraries of different element types (sometimes several hundred). In Creo, we do not have this problem and you can do practically anything with the elements available.

Analysis Methods

For a given model, several different analysis types are possible. For example, in *Structure* the *static* analysis will compute the stresses and deformations within the model, while the *modal* analysis will compute the mode shapes and natural frequencies. *Buckling* analysis will compute the buckling loads on the body. Other analysis methods are available in *Structure* but in this manual, we will only look at static stress and modal analysis. In *Thermal*, the two analysis types are to determine either the steady state temperature distribution, or the transient response of a model given time-varying temperature or heat flux boundary conditions.

Convergence Methods

As discussed above, using the p-code method allows Creo to monitor the solution and modify the polynomial edge order until a solution has been achieved to a specified accuracy. This is implemented with three options:

- **Quick Check** - This actually isn't a convergence method since the model is run only for a single fixed (low, usually 3) polynomial order. **The results of a Quick Check should never be trusted.** What a Quick Check is for is to quickly run the model through the solver in order to pick up any errors that may have been made, for example in the constraints. A quick review of the results will also indicate whether any gross modeling errors have been made and possibly point out potential problem areas in the model.

- **Single Pass Adaptive (SPA)** - More than a Quick Check, but less than a complete convergence run, the single pass adaptive method performs one pass at a low polynomial order, assesses the accuracy of the solution, modifies the p-level of "problem elements", and does a final pass with some elements raised to an order that should provide reasonable results. If the model is very computationally intensive and/or is very well behaved and understood, this can save you a lot of time. The Single Pass Adaptive analysis is the recommended option for most model types. If you suspect problems in the model, or want further verification of convergence, you might want to pick the next option, MPA.

- **Multi-Pass Adaptive (MPA)** - The ultimate in convergence analysis. Multiple "p-loop" passes are made through the solver, with edge orders of "problem elements" being increased by one or two with each pass. This iterative approach continues until either the solution converges to a specified accuracy or the maximum specified edge order (default 6, maximum 9) is reached. Convergence is monitored by watching one or more "measures" or numerical aspects of the solution. Typical measures are the maximum Von Mises stress, or strain energy (as seen in Figure 12). The default measures for a static stress analysis are nodal displacement, local strain energy, and global RMS (root-mean-square) stress. At the conclusion of the run, the convergence measures may be examined. Although it is not strictly necessary, in this manual we will use mostly the MPA option to get a feel for how the analysis is performing.

Design Studies

A *Design Study* is a problem or set of problems that you define for a particular model. When you ultimately press the *Run* button on Creo, what will execute is a design study - it is the top-most level of organization in Creo. There are three types of design studies:

- A **Standard** design study is the most basic and simple. It will include at least one but possibly several analyses (for example a static analysis plus a modal analysis). For this study, you need to specify the geometry, create the elements, assign material properties, set up loads and constraints, determine the analysis and convergence types, and then display and review the final results. The Standard design study is what most people would consider "Finite Element Analysis."
- A **sensitivity** design study can be set up so that results are computed for several different values of designated design variables or material properties. In addition to the standard model, you need to designate the design variables and the range over which you want them to vary. You can use a sensitivity study to determine, for example, which design variables will have the most effect on a particular measure of performance of the design like the maximum stress or total mass.
- Finally, the most powerful design study is an **optimization**. For this, you start with a basic FEA model. You then specify a desired goal (such as minimum mass of the body), geometric constraints (such as dimensions or locations of geometric entities), material constraints (such as maximum allowed stress) and one or more design variables which can vary over specified ranges. Creo will then search through the space of the design variables and determine the best design that satisfies your constraints. Amazing!

A Brief Note about Units

It is crucial to use a consistent set of units throughout your Creo activities. The program itself uses the units brought in with the model from Creo Parametric. There are some tools for dealing with units within the program (like automatic unit conversion). However, the numerical values are provided by you. Thus, if your geometry is created with a particular linear unit like mm or inches in mind, you must make sure that any other data supplied, such as loads (force, pressure) and material properties (density, Young's modulus, and so on) are defined consistently. The built-in material libraries offer properties for common materials in four sets of units (all at room temperature):

> inch - pound - second
> foot - pound - second
> meter - Newton - second
> millimeter - Newton - second

Note that the weight of the material is obtained by multiplying the mass density property by the acceleration of gravity expressed in the appropriate unit system.

If you require or wish to use a different system of units, you can enter your own material properties, but must look after consistency yourself. Table 2-2 outlines the common units in the various systems including how some common results will be reported by Creo. For further information on units, consult the on-line help pages "About Units" and "Unit Conversion Tables."

TABLE 2-2 - Common unit systems in Creo

Quantity	System and Units			
	SI MNS	Metric mm-N-s	English FPS ft-lb-sec	English IPS in-lb-sec
length	m	mm	ft	in
time	s	s	sec	sec
mass	kg	tonne (1000 kg)	slug	lbf-sec^2 / in
density	kg/m^3	tonne/mm^3	slug/ft^3	lbf-sec^2 / in^4
gravity, g	9.81 m/s^2	9810 mm/s^2	32.2 ft/sec^2	386.4 in/sec^2
force	N	N	lbf	lbf
stress, pressure, Young's modulus	N/m^2 = Pa	N/mm^2 = MPa	lbf/ft^2	lbf/in^2 = psi

The units systems for *Thermal* are particularly important (and confusing!). We will leave those until we need them in Chapter 10.

Files and Directories Produced by Creo

Since you will be working in integrated mode in this book, note that your entire simulation model is stored in the Creo part file. You do not need to store a special copy of this. Simulation entities like loads and constraints (plus all other data and settings you define for the model) will appear when you transfer into Creo Simulate from Creo Parametric. It is possible to have different versions of loads and constraints defined within the same model, since the one you want to use is selected as part of the design study. Furthermore, it is possible to suppress model features (usually datum curves) that are created within Creo Simulate.

Creo produces a bewildering array of files and directories. Unless you specify otherwise (or have specified in your default system configuration), all of these will be created in the default working directory. It is therefore wise to create a new subdirectory for each model, make it your working directory, and store the part file there. Locations for temporary and output files can be changed at appropriate points in the program. For example, when you set up to run a design study, you can designate the location for the subdirectory which Creo will create for writing the output files.

The important files and directories are indicated in Table 2-3. In the table, the symbol ① represents the directory specified in the **Run Settings** dialog box for output files, and ② represents the directory specified in the same dialog box for temporary files. Unless the run terminates abnormally, all temporary files are deleted on completion of a run. The names *model*, *study*, and *filename* are supplied by you during execution of the program. Note that many of these files are stored in a binary format and are not readable by normal file editors (or humans!).

TABLE 2-3 - Some Files Produced by Creo

File Type	File/Directory Name	Comments
Model Files	*model*.**mdb** *model*.**mbk**	the **mdb** file contains the last-saved model database. **mbk** is a backup that can be used if the **mdb** file is lost or corrupted
Engine Files	①/*study*/*study*.**mdb**	contains the entire model database at the time a design study is started
	①/*study*/*study*.**cnv** ①/*study*/*study*.**hst** ①/*study*/*study*.**opt** ①/*study*/*study*.**res** ①/*study*/*study*.**rpt**	Engine output files: - convergence information - model updates during optimization - optimization data - measures at each pass - output report for a design study (also accessible with the ***Study Status*** command)
Exchange Files	*filename*.**dxf** *filename*.**igs**	file formats used for import/export of geometry information
Temporary Files	②/*study*.**tmp**/*.**tmp** ②/*study*.**tmp**/*.**bas**	should delete automatically on completion of design study
Results Files	*filename*.**rwd** *filename*.**rwt** *filename*.**grt**	- result window definitions stored with Save in the Result Windows dialog box - result window template definition - graph data report
AutoGEM Files	*model*.**agm**	information about the most recent AutoGEM operation. If the model has not yet been named, this file is **untitled.agm**

On-line Documentation

For further details on any of these functions or operating commands, consult the on-line documentation available with Creo.

Basically, you select ***Help > Creo Parametric Help*** in the **File** pull-down menu. Creo's Help pages are viewed using your default browser.

Once the Help Center has launched, in the left pane select the following:

Simulation > Simulation > Creo Simulate Overview

See Figure 17. Also, check out the Index and Search functions.

PTC Creo Simulate Overview

The Online Help describes how to create a simulation model of the physical nature and real world environment of a part or assembly. You can analyze that model and evaluate the results of the analysis.

Tasks for PTC Creo Simulate	

• Working with the User Interface
  ◦ Working with the User Interface
  ◦ Using the Results User Interface
• Working with the Model
  ◦ Creating a Simulation Model
  ◦ Using Simulation Modeling Techniques
  ◦ Developing a Model in Native Mode
  ◦ Regenerating Models
• Creating Analyses
  ◦ Creating Analyses
  ◦ Managing Analyses
  ◦ Creating Design Studies
• Creating Design Studies
  ◦ Running a Standard Design Study
	• Running Analyses or Design Studies
	  ◦ Checking your Analysis or Design Study before the Run
	  ◦ Running Analyses and Design Studies
	  ◦ Monitoring an Analysis or Design Study Run
	• Reviewing Results
	  ◦ Accessing the Summary Report for the Analysis or Design Study
	  ◦ Evaluating Results
	  ◦ Exporting Results
	• Getting Productive
	  ◦ Designing Process Templates
	  ◦ Changing Configuration Settings
	  ◦ Getting Information on Your Model
	  ◦ Keeping Models Simple

Figure 17 The Simulate *Help Center* main page

Process Guide

The Creo Process Guide is an optional tool to help guide users through the process of creating, running, and analyzing FEA results. It is available in the **Tools** ribbon. Several default guides are available; for example, a guide is available for setting up and running static stress analysis. The guide opens in a separate window on the screen and is used to tell users what information they need to enter for the chosen type of analysis. The guide also offers some explanations for the actions and allows quick links to relevant program commands. Guides can also be customized by users, or written from scratch. This would be useful in order to deploy a specific set of modeling guidelines within an organization. This would make sure all modeling work was carried out consistently, and also offer some assistance to people who are not regular users.

Summary

This chapter has introduced the background to FEA. In particular, the differences between h-code and p-code methods have been discussed. The general procedure involved in performing an analysis was described. Finally, an overview of Creo has been presented to give you a view of the forest before we start looking at the individual trees!

You are strongly urged to have a look at the articles written by Dr. Paul Kurowski that are listed in the References at the end of this chapter. These offer an in-depth look at common errors made in FEA, the concept of convergence, a comparison of h- and p-elements, and more comments on the difference between CAD and FEA.

Another excellent book to read is **<u>Building Better Products with Finite Element Analysis</u>** by Vince Adams and Abraham Askenazi, Onword Press, 1998. This informative, practical, and easy-to-read book was out of print for some time but is now available again.

In the next Chapter, we will start to look at the basic tools within Creo. We will produce a simple model and go through the process of setting up a standard design study for static analysis of a simple 3D solid model. We will also take a first look at the methods for viewing the results of the analysis.

References

The following articles by Dr. Kurowski, although a bit old, contain much useful discussion and may be available on-line at **www.machinedesign.com**:

"Avoiding Pitfalls in FEA," Paul Kurowski, *Machine Design*, November 1994.
"When good engineers deliver bad FEA," Paul Kurowski, *Machine Design*, November, 1995.
"Good Solid Modeling, Bad FEA," Paul Kurowski, *Machine Design*, November, 1996.
"How to find errors in finite-element models," Paul Kurowski & Barna Szabo, *Machine Design*, September, 1997.
"Easily made errors that mar FEA results," Paul Kurowski, *Machine Design*, September, 2001.
"More errors that mar FEA results," Paul Kurowski, *Machine Design*, March, 2002.

The following books, and dozens of others on FEA, may be available in a convenient library:

<u>Building Better Products with Finite Element Analysis</u>, Vince Adams and Abraham Askenazi, Onword Press, 1998.

<u>The Finite Element Method in Mechanical Design</u>, Charles E. Knight, Jr., PWS-Kent, 1993.

<u>A First Course in the Finite Element Method</u>, Daryl Logan, Cengage Learning, 2011.

<u>Practical Stress Analysis with Finite Elements</u>, Bryan MacDonald, Glasnevin Publishing, 2011.

Questions for Review

1. What is the purpose of interpolating polynomials in FEA?
2. What is a "model"? What are some different types of models and how do these relate to the real world?
3. Is it ever possible for a FEA solution to be "exact"? Why or why not?
4. What is the primary source of error when using first-order h-elements for stress analysis?
5. Give an outline of the necessary steps in performing FEA.
6. Why is it probably not a good idea to use a CAD model directly in an FEA solution?
7. What is the "Golden Rule" of FEA?
8. How is convergence of the solution obtained using h-code and p-code methods?
9. Does mesh refinement always yield higher maximum stresses?
10. What is the maximum edge order available in Creo? In the (unlikely) event that the solution will not converge, what needs to be done?
11. What measures are typically used in Creo to monitor convergence?
12. How will error enter into an FEA?
13. What is a *design study*? What types are available in Creo? How is it different from an *analysis*?
14. What are the three methods of convergence analysis? When would each be appropriate?
15. What types of 2D models can be created? In what operating modes? What restrictions are there on 2D models?
16. What types of analyses can be performed on a model?
17. How can you gain access to the on-line help on your system?
18. Compare the advantages and disadvantages of *integrated* and *independent* modes of operation.
19. Is specifying the convergence in Creo considered to be part of the model, analysis, or design study?
20. Where and how do you set up the units for the Creo model?

Exercises

1. Consult a numerical methods textbook and find out what algorithms are used to solve very large linear systems. What effect does round-off error have, and can this be quantified? Are some methods more susceptible to round-off than others?

2. Locate some product brochures for FEM software, and look for the kind of modeling errors discussed in this chapter. Compare the models to the "real thing" and comment on any differences you notice.

3. Another very powerful, computer-based solution method (particularly in fluid mechanics) uses the method of finite differences to solve PDEs. Find out the main ideas behind this method and compare it to FEA.

This page left blank.

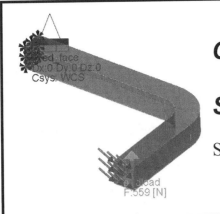

Chapter 3 :

Solid Models - Part 1

Standard Static Analysis

Overview of this Lesson

In this lesson, we will create a very simple solid model and perform a static stress analysis. This will illustrate the common steps involved in all FEA procedures, as outlined in the previous chapter. Along the way, we will encounter the main dialog windows used throughout Creo Simulate for setting program options and will explore some of these. The main steps involved are:

- creation of the model (in Creo Parametric)
- specifying model type
- applying constraints, loads, and material definitions
- defining the analysis and design study
- running the solution
- setting up and showing result displays

After setting up and running this first simple FEA model, we will look at some variations of the finite element mesh and the effect on the results. There are some Questions for Review and Exercises at the end of the chapter, which will give you some practice with the concepts covered and introduce you to some additional options.

This is the longest lesson in this Tutorial, so you might plan to go through this in a couple of sessions. If so, make sure you save your files before taking a break.

Simple Static Analysis of a Solid Part

Solid models are the default type in Creo Simulate. When (most) people think of FEA, they are thinking of or visualizing stress analysis of solid models. These can be treated with a minimum of work and so are a good place to start. You should note, however, that the majority of this tutorial deals with other types of models. Solids are useful for some types of models, but in later chapters we will learn other techniques that are considerably more efficient for some geometries (beams, shells, plates, and so on). In keeping with our "Golden Rule" many of your models will use these idealizations.

The FEA mesh for a solid model is composed of tetrahedral elements. These are like a three sided pyramid; each element has 4 faces (including the bottom), which forms a tetrahedron. The edges and surfaces of each "tet" element can be straight or curved.

Creating the Geometry of the Model

The model we are going to study is a simple L-shaped bar with a square cross section, shown in Figure 1. The bar will be fixed (cantilevered) at the left end, and a uniformly distributed force will be applied to the square face at the other end. We are going to be using this model for the next couple of chapters, so set it up exactly as described below. Also, you may want to think a bit about file management here. For example, because of the number of files created during a run, it is a good idea to keep each model in its own directory to reduce confusion.

Assuming you have Creo running, open the **Folder Browser** and then in the **Common Folders** area, select **Working Directory**. With the cursor in the Browser window and using the RMB (right mouse button) pop-up menu select

> *New Folder*

Call the folder something like **CHAP3**. Open the **Folder Tree**, select the new folder there and, again using the RMB pop-up (this time in the **Folder Tree**), select

> *Set Working Directory*

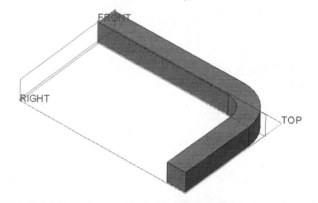

Now close the Browser and create a new solid part. We will use units of mm-N-s (millimeter-Newton-second), which are available with one of the part templates (select the template **mmNs_part_solid**). Call the part *[bar01]*.

The part template has default datum planes and a coordinate system in

Figure 1 Solid model of L-shaped bar

place. Our part consists of a single feature - a sweep. Although there are alternative ways of making this model, this is probably the simplest. There are a couple of ways to make this sweep[1]. We will first create a sketched curve to act as the sweep trajectory. Create a sketched (datum) curve on the **TOP** datum plane. The sketching references are **FRONT** and **RIGHT**. The curve is shown in Figure 2. Notice the dimensioning scheme, which will be important to us in the next chapter. When the curve is complete, make sure it is selected (highlighted in green), and select the *Sweep* tool in the **Shapes** group or pop-up mini toolbar. Observe (and set if necessary) the location of the start point (at the

---

[1] If you are not familiar with sweeps, see Lesson #11 in the Creo Parametric Tutorial, or browse the on-line help.

left end in Figure 2) indicated by the magenta arrow. Use Sketcher (see the RMB menu) to create the square cross section shown in Figure 3 (isometric view), noting that the section will be on the outside of the curved corner - this will also be important later. Make sure you create a solid instead of a surface. Save the part.

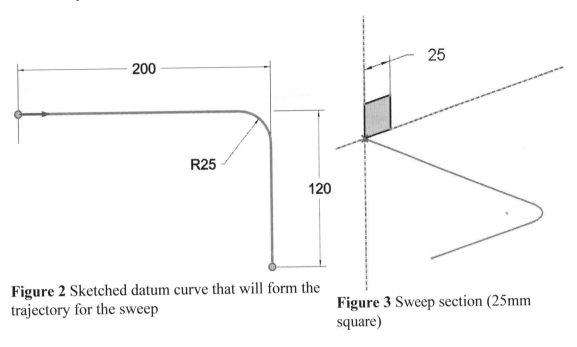

Figure 2 Sketched datum curve that will form the trajectory for the sweep

Figure 3 Sweep section (25mm square)

Setting up the FEA Model

Launching Creo Simulate

With the geometry specified, we are ready to enter Simulate. In the ribbon tabs at the top, select:

Applications > Simulate

The Simulate **Home** ribbon appears[2]. In the **Model Setup** group the two modes of operation are indicated: *Structure* (the default) and *Thermal*. Select the *Model Setup* button in the group. If you do ***not*** have a license for the full version of Creo Simulate, you can check the box beside **Simulate Lite**. This will give you access to all the Simulate functions but restricts your model to no more than 200 surfaces.

_____

[2] Since ribbons are customizable and will adjust to your screen size and settings, your ribbons may be slightly different from the ones shown below.

---

* * * * IMPORTANT * * * *

The check box *FEM Mode* allows you to do model pre-processing *only*. With this checked you can apply loads, constraints, define materials, and create the mesh, BUT you cannot run the solver. This mode of operation allows you to create models and produce correctly formatted input files for other commercial FEA solver programs. This is handy because some solvers do not have very easy-to-use preprocessor interfaces, so you can use Creo to set up the model much easier. **If you are going to use the Creo solver, *FEM Mode* must be left unchecked.**

Click the *Advanced* button. This shows the other model types we will be investigating through these lessons. The default type, as mentioned above, is a 3D solid. The *Default Interface* pull-down list contains the options for how Creo will treat mating parts in an assembly. We will investigate some of these options later. This will obviously have no effect for single parts. Accept all the defaults in the **Model Setup** window, then *OK*. If you want to change the model type in the future, you can select *Model Setup* any time. If you have accidentally chosen the wrong part units, or want to check on the units in an inherited part, in the **Setup** pulldown select *Units*. This opens the **Units Manager** window. In our case, make sure this is set to "millimeter Newton second (mmNs)".

The part display is essentially the same as the Creo display. Notice the addition of a green coordinate system labeled **WCS** (visible if *Csys Display* is turned on). This is the *World Coordinate System*, created by default by Creo. An FEA model can have several coordinate systems (cartesian, cylindrical, or spherical). The currently active system is highlighted in green. Note the directions of the X, Y, and Z axes - these are important to specify directions of loads and constraints. You can turn off all the datum displays (use the Graphics toolbar), as these will not be needed any further. The orientation of the WCS is still shown at the lower right - this is handy when setting up directions for loads and constraints.

Take a few minutes to browse around the remaining ribbons shown in Figure 4. As you develop your model in Creo, you will generally work left to right across the **Home** ribbon using the group commands for **Loads**, **Constraints**, **Materials**, and finally **Run**. The commands in the other ribbons are for the following:

> **Refine Model** - creation of idealizations like shells and beams, special connections, surface and volume region definitions, datum creation, and mesh controls.
> **Inspect** - to obtain information about the model.
> **Tools** - commands for querying the model or accessing the Process Guide.
> **View** - commands for modifying model appearance and display properties.

Note that the datum creation group is still with us (in the **Refine Model** ribbon). This indicates that it is possible to create these features while we are in Creo if the need arises. If we do, they will be saved with the model but only active (and visible in the model tree) when we are in Creo Simulate; the same is true for everything we create.

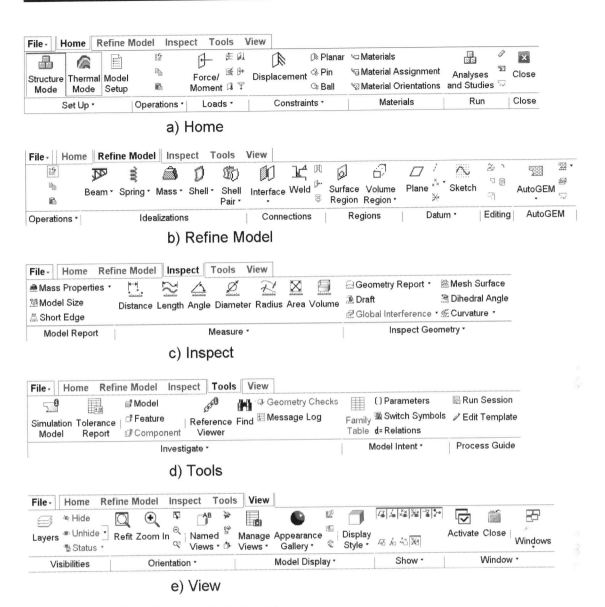

a) Home

b) Refine Model

c) Inspect

d) Tools

e) View

Figure 4 The Creo Simulate default ribbons

Check the preselection filter options in the pull-up list at the bottom right of the screen. This filter is to help you when selecting simulation entities on the screen, and works in the same way as the preselection filter in Creo Parametric.

Applying the Constraints

We are going to constrain the left end of the bar against all motion, since the bar is cantilevered out from a support. This constraint applies to the entire square surface at the end. Since this will be a fixed planar surface, there are a couple of ways to set up this constraint. We will look first at the most general procedure. Select

Home { Constraints } : Displacement

A new dialog window opens (Figure 5). In the top data field, enter a name *[fixed_face]*. Individual constraints are stored as members of constraint sets. The default name of the set is **ConstraintSet1**. In the **References** list, note that the default is *Surfaces.* We will see later why this is a useful default! With the **Surfaces** collector active (red circle), pick on the desired surface on the model. You may want to spin the model. The surface will highlight in green when selected[3]. Notice that the default coordinate system is WCS. If you wanted a different system (if it existed!), you could select that here. At the bottom of the **Constraint** window, there are six possible constraints - three translation and three rotation constraints. Each constraint has four buttons. The buttons (from left to right) are for

- **Free**
- **Fixed**
- **Prescribed** (specified displacement)
- **Function of coordinates**

The default setting is **Fixed** for all translation constraints. This is exactly what we want here for the translation of this surface. Any element nodes that lie on this surface will be prevented from moving in any direction. If you pick a **Prescribed** constraint, you would enter the specified displacement.

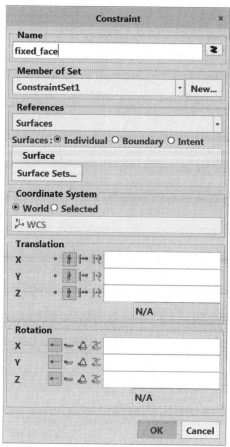

Figure 5 Displacement constraint dialog window

IMPORTANT POINT

 Rotational constraints are irrelevant for solid models. The FEA solution of a model using solid elements involves only the translational degrees of freedom of the nodes. Thus, rotation of a node can be neither specified, constrained, nor computed since it does not mathematically exist!

The completed constraint dialog window is shown in Figure 5. Select *OK* or middle click. The surface will initially appear in green.

If you click anywhere else in the graphics window, the constraint display turns purple. If you mouse over any of the markers, the following will appear: the constraint name ("fixed_face"), some information about the degrees of freedom (Dx, Dy, Dz), with 0 meaning "fixed", and the coordinate system these relate to. If you zoom in on the surface,

---

[3] If you accidentally middle-click here, the dialog window will be accepted. Select the constraint symbol in the graphics window, and use the RMB pop-up menu to select *Edit Definition*.

a large triangular icon appears (Figure 6). This is the constraint icon[4]. The constraints are indicated by the fill pattern of the six boxes across the bottom of this icon (top row: Dx, Dy, Dz; bottom row: Rx, Ry, Rz). The translation constraints are currently all fixed, so all boxes are filled in. The lower row of empty boxes indicate that rotation constraints are not applied. To change the display of the constraint, use the *Simulation Display* tool in the Graphics toolbar. We will explore that a bit later. If you need to change the constraint settings, select the constraint icon then in the mini toolbar select *Edit Definition*.

Figure 6 Constraint icon

Applying the Loads

The constraint we just applied used an action/object command form. Let's try object/action for the load. Select the square face at the other end of the bar. In the pop-up mini toolbar select

Force/Moment

This brings up the dialog window shown in Figure 7. Enter a name *[endload]*. The load is a member of **LoadSet1**. Once again the default reference type is **Surfaces**. The collector already contains the surface at the free end of the bar. The load will be defined relative to WCS. Select the *Advanced* button, and check out the options available for the load distribution. The default *Total Load* and *Uniform* settings are exactly what we want. The areas at the bottom of the dialog window are for entering the applied forces and moments on the designated surface. These can either use components or (see the pull-down list) a magnitude and direction. Select *Components* and enter the values shown in Figure 7. Note that the units are displayed. Recall the directions of the WCS. Click the *Preview* button to see the applied load on the model. Finally, select *OK*.

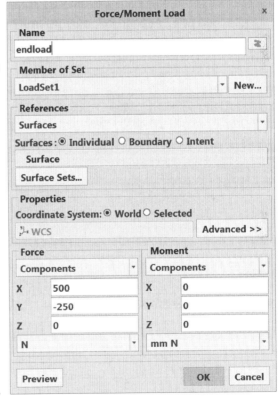

Figure 7 Force/Moment dialog

---

[4]It will be easier to see some of the labels if you go to wireframe, or use transparency in the model appearance.

The model should now appear similar to Figure 8.

There are many variations available for the display of the loads and constraints on the model. Let's explore some view options. In the **Graphics** toolbar, select

Simulation Display

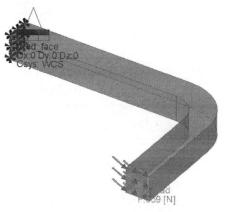

Figure 8 Model with load and constraint applied

This brings up the dialog window shown in Figure 9 which allows us to control the appearance of the display. Experiment with the various options by checking them and using the **Preview** button. For example, try out the **Display Names** and **Load Value** options for the icons. These displays are handy for documenting models - a hard copy of this display is a useful image to archive. Change the Load Arrows display to **Arrow Tails Touching**. When your models get more complicated, it is useful to have these controls over the display to relieve the screen clutter. Check out the other tabs in this dialog window. Leave the window as shown in Figure 9 and select **OK**. The model should appear as in Figure 10. The single bold orange up-arrow icon indicates the load on the end surface. Notice that it is not in the direction of the applied load - it is merely an iconic symbol similar to the constraint icon.

Figure 9 Setting up the simulation display settings

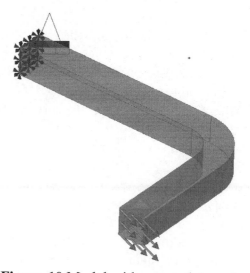

Figure 10 Model with constraints and load defined

Specifying the Material

Now we have to tell Creo what the bar is made of. This is a two-step process: bringing a material definition into the model, then assigning it to the part. In the **Materials** group,

select

> ### *Materials*

The **Materials** dialog window opens (Figure 11). In the list at the left, select the directory **Legacy-Materials**, then browse down and click on **steel.mtl**. This transfers the material definition to the model (but we have not yet assigned the material to the part). Remember that a model (of an assembly, for example) can have several materials in it, each assigned to one or more different components. You can see the numerical values of the material properties in the panel at the right. If you don't like the units in their current format (like tonne/mm³ for density!) or wish to change a numerical value, you can double click the entry in the model area at the bottom. This opens another window where all the properties can be modified. For example, select from a pull-down unit list for density to get something more familiar (like kg/m³). *Cancel* this window and select *OK* in the **Materials** dialog window.

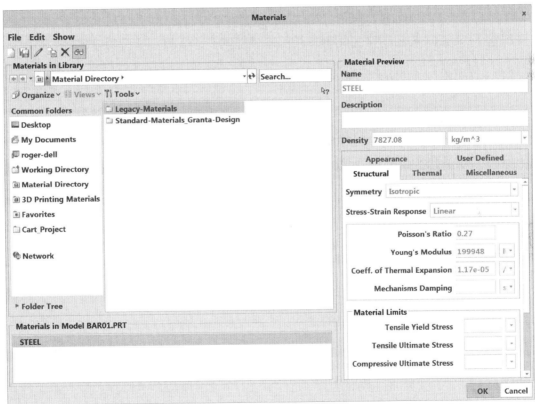

Figure 11 The dialog window to bring material definitions into the model

In the **Materials** group select

> ### *Material Assignment*

The bar is picked automatically since it is the only component in session. Notice the *More* button beside the listed material; this takes us back to the previous material selection dialog window[5]. To apply the current material to the part, just select *OK*. A new icon appears on the model to indicate the material definition is applied.

Our model definition is now complete. Think back to what we have created so far: geometry, model type, constraints, loads, and material properties. These are the essential elements of the FEA model. Check out the model tree to see the information stored there. You should save the model at this point.

Setting up the Analysis

We can now tell Creo what to do with this model. In the **Run** group, select

Analyses and Studies

which brings up the **Analyses and Design Studies** dialog window (Figure 12). This window gives you access to all the commands for setting up how you want the model analyzed, how you want convergence to be monitored, where to put the results, and so on. Collectively, these settings define a *study*. The simplest study is a simple static analysis, which is what we will do here. We will get to sensitivity studies and optimization a bit later. This window is also where you actually give the command to run the study - so we are getting close to that!

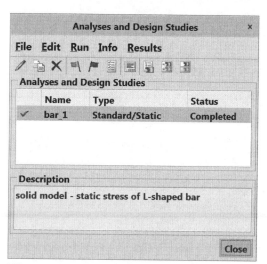

Figure 12 The main dialog window for defining and running analyses and design studies

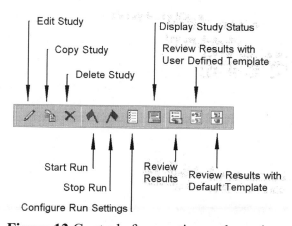

Figure 13 Controls for creating and running analyses and design studies

Most commands and functions in this window have several possible entry points - the pull-down menus at the top or the row of buttons in the middle (see Figure 13). The studies we define will be listed in the central pane just below these buttons. Obviously, we can have more than one study defined at any given time. Selecting a study here and

---

[5] HINT: As a shortcut, for a part model we could have bypassed the previous step of bringing the material into the model.

using the RMB pop-up gives you yet another way to launch commands.

In the pull-down menu, select

File > New Static

This opens the **Static Analysis Definition**
dialog window shown in Figure 14. Enter
a name for the analysis *[bar_1]* at the top
of this dialog.

IMPORTANT

The analysis name you just entered will
be the name of a subdirectory containing
all your result files for this analysis. This
name will be important later. It is the
study name indicated in Table 2-3 of
lesson 2.

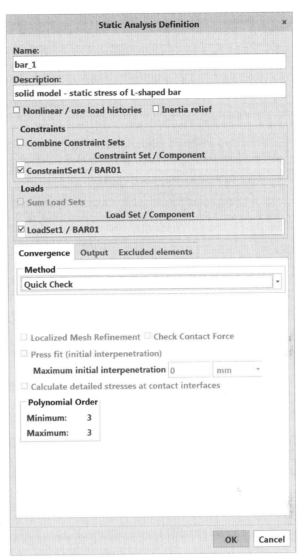

Figure 14 Creating a Static analysis

You can enter a short description for the
analysis in the Description box. For a
static stress analysis we need to
specify/select the constraint and load sets.
These are the ones we created above
(default names **ConstraintSet1** and
LoadSet1) and are entered automatically.
We must also specify what type of
convergence analysis we want to use,
available in the **Method** pull-down list
under the *Convergence* tab. The types of
convergence analysis available here were
discussed in Chapter 2. The first run of a
model should always be a *Quick Check*.
This does a "complete" analysis of the model at a low (usually 3) edge order. The results
of a Quick Check should never be trusted, since no convergence or accuracy information
is available. The purpose of the Quick Check is to make sure the model is solvable. For
example, if we have improperly constrained the model, the solver will intercept this and
give us an appropriate error message. The completed dialog window is shown in Figure
14. Select *OK*. Back in the **Analyses and Design Studies** window, you can see the
analysis **bar_1**, type **Standard/Static** as in Figure 12.

To see if we have any gross modeling errors, in the pull-down menu select

Info > Check Model

This can trap some types of errors in the model (for example, no materials specified).
We should be informed that there are no errors in the model. If errors have been

detected, go back through what we've just done and make sure all the steps were completed. See what happens if you **Suppress** our **fixed_face** constraint in the model tree and then run **Check Model**. What happens? **Resume** the constraint, and we are on to the next step.

Running the Analysis

This is what we've been waiting for! Still in the **Analyses and Design Studies** window, in the pull-down menu select (or use the toolbar button **Configure Run Settings**)

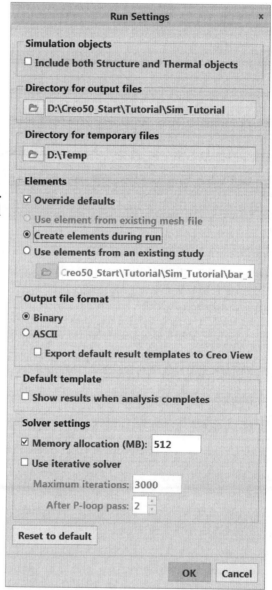

Run > Settings

This opens the window shown in Figure 15. Here is where we can specify where Creo should direct all its temporary and output files. If all goes well, the temporary files are deleted automatically after a successful run. You normally don't bother with them, so sending them to a "trash" directory is all right. Be aware that if either of these directories is on another computer on a network then you will seriously slow down the operation of Creo. The default output file directory (unless specified in a special configuration file for your system) is the current working directory. Set this to wherever is convenient for you. We will discuss the Elements options a bit later. There is an option (default OFF) for showing results as soon as the run finishes using some default templates. Leave this off for now - we will discuss result templates later. The dialog window also has an option for specifying how much memory should be allocated to the run. In this tutorial, our models will be small enough that the default is satisfactory. If this allocation is too small, Creo will have to page out to the hard disk, which slows it down. A good rule of thumb for optimum performance

Figure 15 Specifying the run settings

of really large models is to set the **Memory Allocation** to half the physical memory in your computer. Accept the **Run Settings** dialog with **OK**.

Back in the **Analyses and Design Studies** window, select

Run > Start (or use the green flag in the toolbar)

You are asked if you want to run the interactive diagnostics. This can help you find some modeling errors if things go wrong. Select *Yes*. In the message window you will see[6] "The design study has started." Creo goes through the process of automatically generating the element mesh, formulating the equations, and then solving them. Meanwhile, a new Diagnostics window will open that indicates various steps of the solution and some brief error messages if any. That shouldn't happen with this model, so close this window. You can also follow progress by selecting the *Display Study Status* button in the toolbar. The **Run Status** window appears (actually, just an extension of the previous Diagnostics window) with a scroll bar on the right. The **Summary** tab displays a report file that is saved after the run in the output directory (*bar_1.rpt*). Use the scroll bar to go to the top of the report.

As indicated in the report file, the automatic mesh generator, AutoGEM, creates 20 or 21

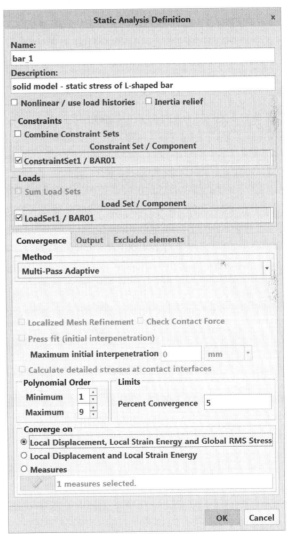

solid elements. A lot of other information is also given. Browse through this to see what is there. Near the bottom of the report, in the Measures area, you might note that the maximum displacement magnitude (**max_disp_mag**) is about 0.54 mm, and the maximum Von Mises stress (**max_stress_vm**) is about 43 MPa. We will compare these results to our "converged" results a bit later. Most importantly, the Quick Check analysis completes with no errors - our model is OK. In the **Run Status** window, select *Close*.

With an error-free Quick Check behind us, we want to change the convergence analysis to Multi-Pass Adaptive. As discussed in the previous chapter, this performs an iterative process. The model is analyzed first using low order elements. An internal algorithm estimates the error in the solution and increases the polynomial order of the offending elements. This process continues until the estimated error is less than the specified tolerance.

Select the analysis **bar_1**, and in the RMB pop-up menu select *Edit*. Now, pick the *Convergence* tab, and in the pull-down Method list, select *Multi-Pass Adaptive*.

Figure 16 Editing the analysis convergence settings

_____

[6] Actually, for this model, by the time you have read this, the run is probably finished!

Set the maximum polynomial order to **9** (the maximum possible), the percent convergence to **5**, and note the radio button beside **Local Displacement, Local Strain Energy and Global RMS Stress**. These settings are shown in Figure 16. We are asking for the convergence iterations (the "passes") to proceed until all of these quantities change by less than 5% between iterations. We will let the element polynomial orders go as high as 9 if necessary (note the default was 6). Accept the settings with *OK*.

Now open the ***Run Settings*** dialog window again. Note the default selection of "Use elements from an existing study" and the location of the study. We don't need to generate the elements again (on a bigger model, this might be a lengthy process), since our model has not changed. If you wanted to create a new mesh, you would check the box beside **Override Defaults** and **Create Elements During Run**. We do not want to do this now. Close this window and select ***Start Run***. Creo detects the output files from the previous run (the QuickCheck). Select *Yes* to delete these files, then *Yes* to see the **Interactive Diagnostics**. The MPA analysis will start up. The run will take a few seconds more than last time. On completion, the diagnostics window will appear as in Figure 17.

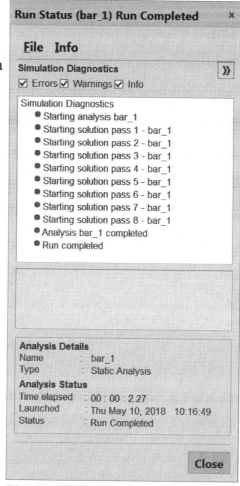

Figure 17 The Diagnostics window.

The run converges[7] on about pass 7 or 8. Open up the ***Study Status*** window to check the maximum edge order. A great deal of data is presented about the converging solution. We will be plotting some of this in a minute or two. The maximum displacement magnitude is 0.56 (mm) and the maximum Von Mises stress is 44 (MPa). Compare these to the results of the Quick Check obtained above. It is very unusual for the results to be so close together. We will see why this occurred a bit later. Before we leave, take note of the elapsed time and CPU time reported for this run at the bottom of the report - we will compare these with some runs we'll make a bit later using a different mesh. *Close* the **Run Status** report window and the **Design Studies** window.

---

[7] As mentioned previously, convergence behavior will vary somewhat depending on how the model was created and the software version (or build). Sometimes small changes in the meshing or solution engine will manifest as differences in the mesh or solution behavior.

Displaying the Results

Creating Result Window Definitions

As mentioned in an earlier lesson, FEA produces an enormous amount of output data. The results directory for the analysis **bar_1** just completed contains 26 or so files and is around 1.3 MB in size for this very simple model. For more complex models, it is not uncommon to have results directories up to a hundred megabytes in size (or more). Reviewing and interpreting this data is best done graphically.

The FEA results can be presented in many ways. Primarily we are interested in various views of the deformation (either displacement components, total displacement magnitude, or material strain components), the stresses in the model (components of normal and/or shear stress, von Mises stress), and data that illustrates the convergence behavior of the analysis. For modal analysis, we can display the mode shapes; we can also produce shear and bending moment diagrams for beam elements. Some of this data represents the primary solution variables (in the case of stress analysis, this is the deformation) while other data (like the material strain and stress) is derived from the primary variables.

In Creo, we display these FEA results in *result windows*. We can display these individually or several simultaneously on the screen. Each result window has a unique name and the contents are determined by a *result window definition*. The set of result window definitions can be saved in an *rwd* file for later use (i.e. the window definitions don't have to be created again if the analysis is opened up later). Default settings for background colors, text fonts, and so on can also be modified. Also, results from several models can be shown on the screen simultaneously, each in a different result window. This is handy for comparing results from different models or analyses.

Let's see how all this works. The procedure for displaying the results is as follows:

- create and name the new result window
- identify the location (i.e. the directory) of the result data for the window
- specify the data to be plotted in the window
- select display options for the data
- show the result window
- use optional controls to manipulate the result display in each window

This process is repeated for each result window (with some shortcuts available by copying existing definitions). Some of these steps are automatic. To start this process, select

Home { Run } : Results

There is also a short cut to this command in the **Design Study** dialog, either in the pull-down menus or in the toolbar. You can experiment with these later.

If you are asked to save the current model, select *No*, since that has been done automatically already. A new window opens up on top of the Creo main window. The new untitled window contains some new ribbons - see Figure 18. Most of these are grayed out at this time but will become active when the first window is defined.

				Copy		Model Max	View Min			Display Element IDs
New	New from	New	Edit	Delete	Dynamic	Model Min	Clear Tags	Measures	Linearized	Display Node IDs
	Template	Default			Query	View Max	Clear All Tags		Stress	Display Result Values
	Window Definition					Query			Report	

a) Home

		Zoom out	Spin Center			Transparent	Edit			Tile	Reorder
Refit	Zoom in	Repaint	Saved Orientations	Shaded	Continuous	Exploded	New	Delete	Slow——Fast	Cascade	Hide
		Saved Views	Default View		Tone	Overlay				Swap	Show
	Orientation				Appearance		Capping & Cutting Surfs	Animation		Window	

b) View

Edit	Reset		Tie Quantity	Untie		Edit	Show Annotations
Tie	Contour Label Density	Edit	Tie Location	New	Delete		
Untie			Tie Both		Delete All		
	Legend		Graph		Annotation		

c) Format

Figure 18 Ribbons for creating and manipulating result windows

The first window we will create contains a color fringe plot of the Von Mises stress. If a dialog window does not appear as in Figure 19[8], select the *Open* button on the left in the **Home** ribbon and then choose the design study **bar_1** and *Open*.

In the dialog box that opens (Figure 19), enter a name *[vm]* for this result window. Now we must tell Creo where the data is (the design study or analysis results). Depending on how you got to this dialog window, some of the information may already be filled in. If so, skip to the next paragraph; if not, read on. You may recall that we named our analysis **bar_1** and in the **Run Settings** dialog we told Creo where to put all the output files. At that location, Creo has created a subdirectory called **bar_1** (i.e. the same name as the study). Click on the button below **Design Study**, navigate to this directory location and select it, then *Open*. All you need to specify is the top subdirectory named **bar_1** - Creo will find all the files it needs inside there. The dialog window expands for us to define the contents for the result window.

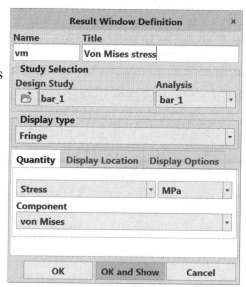

Figure 19 Defining contents of a result window

---

[8] This dialog opens automatically if you launch the *Results* command from within the **Design Study** dialog window, since the design study is already known then.

In the title field at the top, enter a window title like *[Von Mises Stress]*. The defaults are *Display Type(Fringe)* and *Quantity(Stress)*. The default stress is **Von Mises**. Click on the **Display Options** tab and select the box beside **Show Element Edges** (Figure 20). This will let us see the mesh. Now select the *OK and Show* button at the bottom of the dialog window.

The display now shows the fringe plot of the von Mises stress, and most of the remaining ribbon buttons are now live. You can spin/zoom/pan this display using the mouse buttons. You can also manipulate the display using commands in the *View* ribbon or RMB pop-up menu. We will return to this display in a while to explore other options. In particular, the colors might appear a bit murky - we need to fix that (later!).

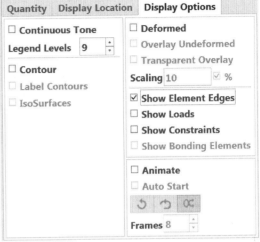

Figure 20 Result window display options (see also Figure 19)

Now, we want to create a number of other result window definitions. We have two options here:

☞ If we use *Open* again, we will be asked to specify the results directory again. This is handy if you want to display results from two different models or analyses for comparison.

Or

☞ If you use the *Copy* toolbar icon, the result directory and window definition from the current window will be used automatically, saving you several steps and mouse clicks. All you have to do is supply the new window name, and change the desired parts of the definition.

You should experiment with both these options later. For now, since the next window we want will show an animation of the deformation, that is the same results directory, select the *Copy* toolbar button. In the new dialog window (Figure 21), enter a name *[def]*. Change the title to *[Deformation]* and set the following options (might as well go for broke here!):

Display Type (Model)
Quantity(Displacement, Magnitude)
Display Options
Shade Surfaces | Deformed | Transparent Overlay
Scaling 10% | Show Element Edges | Animate | Auto Start | Reverse
Frames 24

See Figure 22. One item to note here is that the number of animation frames must be a multiple of 4. Too few frames will be jerky, too many will be too slow. You can come back later to experiment with other options in this dialog window. Accept the definition with **OK**. We will show this result window in a minute. Note that we can define a result window without displaying it - this makes sense if you have a complicated model with many result windows and you don't want to see them all at once.

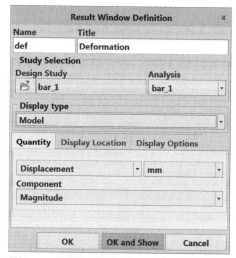

Figure 21 Defining a second result window to show deformation

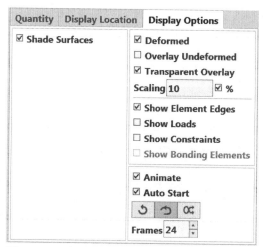

Figure 22 Display options for second result window (see Figure 21)

As promised before, we are going to plot some data to illustrate the convergence process that occurred during the multi-pass adaptive run. We are interested in the stress, deformation, and strain energy. Some key values ("measures") for these quantities are computed automatically and stored for each pass during the run.

Select the **Copy** toolbar button again and enter a name for the new window, *[convm]*. Change the title to *[Von Mises Convergence]* and set the following options. See Figure 23.

> *Display type(Graph)*
> *Quantity(Measure)*

Pick the button under the **Quantity** pull-down list, scroll down the list in the **Measures** window and select

> *max_stress_vm | OK*
> *OK and Show*

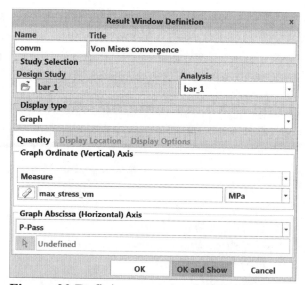

Figure 23 Defining a result window to show convergence of a measure

The screen will now display two result windows side by side (von Mises stress and stress convergence). See Figure 24. The format of these windows has been modified a bit using commands we will see later.

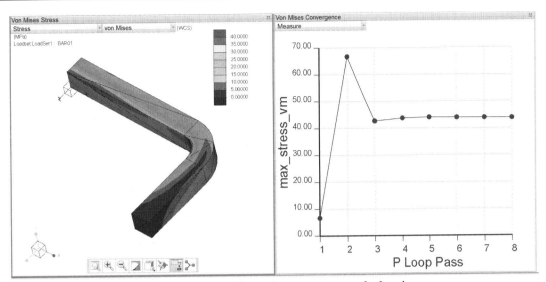

Figure 24 Von Mises fringe plot and MPA convergence behavior

One of these windows is "active", indicated by the darkened border. Click on the convergence graph window so that it is active. Now we can copy this definition to another new window, keeping many of the same definition settings. Select *Copy* to create another window called *[condef]*. The title of this will be *[Deformation Convergence]*. The measure we want to plot is *max_disp_mag* (maximum displacement magnitude) - use the button under the **Quantity** list box to get it. See Figure 25. Accept the dialog with *OK*.

Finally, *Copy* the convergence window definition to another window called *[constr]*. The title of the window is *[Strain Energy Convergence]*. The measure we want to plot here is at the bottom of the list, *strain_energy*. Accept this definition with *OK*.

We have now defined five result windows, two of which are currently displayed. To see a list of the defined result windows, in the **View{Window}** group, select

Show

This will open the window shown in Figure 26. To view the various result windows, pick their name in the list, and then select the *OK* button. Windows can be shown one at a time or several together.

To change any definition settings for a result window, you must first display the window then make it active by clicking on it. To change the window definition, select *Edit* in the **Home** ribbon, or even easier, in the RMB pop-up in the graphics window.

Let's see what can be done with the individual window displays.

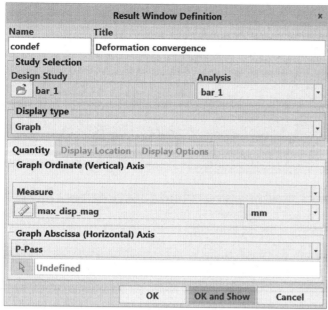

Figure 25 Defining a window to show displacement convergence

Figure 26 List of defined result windows

Showing the Result Windows

We'll start with the deformation window. Highlight only the entry **def** in the list of Figure 26 and then select *OK*. A view of the model appears (wireframe or shaded depending on previous options) animating the deformation. Recall we set up AutoStart in the result definition dialog (Figure 22). All the element edges are shown. Some information is given at the top of the window. You can spin/zoom/pan the model using the usual mouse buttons.

Use the animation control buttons in the **View** ribbon (or RMB pop-up) to start and stop the animation, and to step forward and backward one frame at a time. Notice that the deformation has been magnified by a large scale factor (listed at the top left) for us to see it. Remember that the actual maximum displacement is only about 0.5mm. This is only 1/50th of the bar's cross section width - only a few pixels on the screen. In order to "see" the deformation it must be magnified

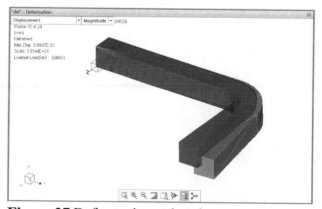

Figure 27 Deformation animation

quite a bit - this is common practice in FEA. The maximum deformation is shown in Figure 27.

The most important information in an animation like this is whether the deformation is consistent with our applied boundary conditions and loads (or at least what we think they are!). If you make a serious mistake with these (like loads in the wrong direction), it will usually be obvious when you watch this animation. It is useful to produce this animation immediately after running the Quick Check analysis to verify the loads and constraints on the model. Check out the ***Overlay***, ***Transparent***, ***Continuous***, and ***Shaded*** commands to see which combinations work or not.

Stop the animation, open the **Display Result Window** with ***Show*** (Figure 26) and select just the window **vm** in the list, then select ***OK***.

A colored fringe plot of the part appears with a legend at the top right. The colors correspond to different levels of the von Mises stress. If your colors appear "murky", go to the **View** ribbon and deselect the ***Shaded*** option. In the RMB pop-up select ***Edit*** then in the **Display Options** tab turn on **Show Loads** and **Show Constraints**. The display is shown in Figure 28. When you are finished exploring this display, let's experiment with some of the options available for viewing the Von Mises stress fringe plot.

Open the ***Display Options*** tab again (using the RMB menu ***Edit*** command) and select **Continuous Tone**. Turn off the element edges, loads, and constraints. Accept this window with ***OK and Show*** - it should look like Figure 29. This is a very pretty picture (the kind the vendors like to put in glossy sales literature!), but it is difficult to be precise about the color when interpreting the stress levels. This is an example of the kind of post-processing "magic" that can be done. Remember that both Figures 28 and 29 are derived from the same data. Which figure do you think most people would think is "more accurate"? Would you agree?

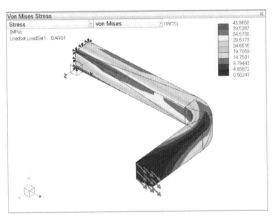

Figure 28 Von Mises stress fringe plot (8 levels, with element edges)

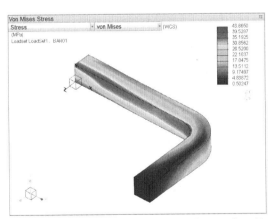

Figure 29 Von Mises stress fringe plot (9 levels, continuous tone, no edges)

Now select

Format { Legend }: Edit

This shows the minimum and maximum values in the legend scale. This dialog is basically used to set the number of levels (currently 9) and the color spectrum. Go ahead and experiment with these to see their effect on the display. In this dialog you can also set the visibility of the legend, and the minimum and maximum values to be used. The legend values are linearly distributed.

Suppose you only want to see values that exceed a certain limit (like a certain percentage of the yield stress for the material). In the edit legend window, set the minimum value to 30, then *Preview*. Click on each of the two lowest colored boxes and change the color to your background color (white?), then *Preview*. Only stresses greater than the fringe minimum are displayed. When you are finished, set the minimum value to **0** and the maximum to **40** (this gives us "nice" levels in the legend), and return to 9 levels with the spectrum *Structural*.

Edit the window definition for **vm** once again using the RMB pop-up. In the **Display Options** tab, turn on *Deformed* and *Animate* (24 frames). Select the *Reverse* button. Now *OK and Show* this new definition. You will see a shaded image of the bar, with animated stress fringes. Pretty impressive! Use the animation controls in the **View** ribbon to start/stop, step, and change speed of the animation.

Remember that this animation is not "real" since it is based on a single static solution which is being linearly interpolated between the zero-load (zero deflection) and full-load (maximum deflection). Such an interpolation is possible only if the problem is fully linear (materials and geometry). For a non-linear problem, the deformation might not appear this way as the load is being applied.

One thing we'd like to know is where does the maximum stress occur? Stop the animation and *Edit* the definition of **vm** again. Leave it as a fringe display, but turn off the deformation and animation. Turn element edge display back on and *Show* the window. In the **Home** ribbon, select

          Model Max (or use the RMB pop-up)

This locates the location (with a small triangle and XYZ coordinates) and gives the value for the maximum stress. Note the difference between this and *View Max*, which shows the point and value of the maximum stress in the current window. To see that these are different, zoom in to get a closeup of the bar near the constrained end, then select *View Max* and *Model Max* - the two are now different. Now select (in the ribbon or RMB)

          Dynamic Query

Put the mouse cursor over the model and move the mouse around. The stress level at the cursor location is shown in the small box to the right of the display. If you click the left mouse button, a marker and label will be put on the display. When you are finished, select *Close* in the **Query** window (or middle click). To remove all the tags (labels), in the **Home** ribbon or RMB pop-up select

          Clear All Tags

Another view of the data can be obtained using a cutting surface. In the **View** ribbon, go to the **Capping & Cutting Surfs** group and select

New (or *New Cutting/Capping Surfs* in the RMB pop-up)

We want to cut the model along a horizontal plane halfway between the upper and lower surfaces. This is parallel to the WCS XZ plane[9]. The cutting surface definition is shown in Figure 30. To see the surface, select *Apply*. The model is now sliced on the cutting plane, and we can see the stress levels on that plane.

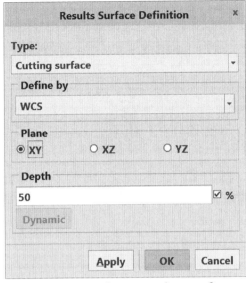

Figure 30 Creating a cutting surface

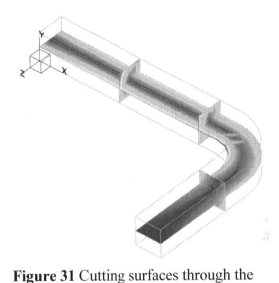

Figure 31 Cutting surfaces through the model

In the **Results Surface Definition** window, select *Dynamic*. Now click and drag the left mouse button in the display window. You can move the cutting surface up and down in the model. When you are finished, middle click, then *Cancel*. Note that you can create more than one cutting surface, as shown in Figure 31.

To remove cutting and capping surfaces, look for the *Delete* command in the same ribbon group.

A *Capping Surface* is similar to a cutting surface, except that the material is removed on one side only. Try setting up a capping surface parallel to the YZ plane. You can only have one capping surface at a time.

Now we will look at the convergence plots. You can highlight all three windows at once in the **Show** list and then select *OK*. The three graphs are shown in Figure 32.

---

[9] Note that you can use other coordinate systems if they are defined.

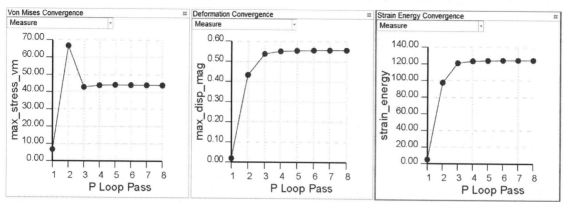

Figure 32 MPA Convergence graphs for (left) **Max VM Stress**, (middle) **Max Deformation**, and (right) **Total Strain Energy**

We observe in the convergence plots that, as far as these measures are concerned, the solution essentially was unchanged after the 3rd pass. This explains why our results are so similar to the Quick Check performed earlier (which used order 3 elements throughout). However, at least one measure being monitored during convergence (but not plotted here) did not meet the 5% criterion that we set. As a result, there must be at least one edge in the model that has been bumped all the way up to 8th order.

To see the p-levels in the model, create a new window called *[plev]* and select

> *Display Type (Fringe)*
> *Quantity (P-level)*

Show this result window and you will see several edges were 8th order. Curiously, these were not in the region of maximum stress. These elements do, however, span a considerable distance in the model - all the way from the constrained surface to the filleted corner. In order to capture the shape of the variation somewhere along these edges, evidently the polynomial orders needed to be quite high.

Nonetheless, this solution appears to be very well behaved. It is certainly much better behaved than other models you are going to see. Note that we probably could easily limit the maximum edge order to 3 or 4 (thus significantly reducing our solution CPU time) and still obtain very similar results for this model. This is exactly what we will do in the next chapter when performing a sensitivity study and optimization of this bar.

The format of the result windows containing graphs (like those above) can be changed quite easily. Show one of the convergence graphs and then select

> *Format { Graph } : Edit*

(Note that the Edit command in the RMB pop-up takes you to the result window editor.)

This opens the **Graph Window Options** dialog window. This dialog has controls for line color and width, text font (style, size, color), axis labels, data symbols, and so on. There is lots to experiment with here.

Before leaving the result window, check out the *File > Save As* command and look in the **Type** drop-down list at the lower right. This provides tools for archiving the graph data (look for *.grt files), or sending it to other programs (like a spreadsheet).

We are finished looking at results for now. Before we leave we should store all our result window definitions so that we don't have to recreate them if we come back to this model later. Use *File > Save As > Save a Backup*. This opens a dialog window where you can specify the name and location of the *rwd* file (result window definition) containing all the currently defined result window settings for this model. Name this file [*bar_1results*] (rwd is added automatically) and store it in the same directory as the model.

If you want to use (more-or-less) the identical result window definitions for other models (which we do!), select

File > Save As > Save a Backup

In the **Save As** dialog window, in the **Type** dropdown list at the lower right select *Simulate Results Template*, and enter a name and location (might as well use the default Creo working directory) for the template file - something like "*MPA_standard*". The extension "rwt" will be added automatically by the system. All the definitions of the currently defined result windows will be stored. Do not save units, legend values, deformed scale, or model orientation with the template for now (although it is nice to know that can be done). Where *.rwd files are specific to a given model, *.rwt files can be used for multiple models, with some restrictions. For example, windows such as convergence graphs that show results for an MPA analysis are meaningless for a QuickCheck or SPA. You might like to make a version of a result window template specifically for QuickCheck runs, that would contain only an animated view of the model and the Von Mises stress.

To see if your rwt file is working, in the **Home{Window Definition}** group, select *Close All* to remove all current result windows. Then select

Open From Template

.You must first identify the source of the data to be shown, i.e. the design study. Select **bar_1** in the current working directory. In the small dialog window that opens, the design study and analysis are listed (only one analysis at this time). Select the button under **Template Selection** and locate the rwt file we made above (something like *MPA_standard*). Then *Open > OK and Show*. All six result windows are created and displayed. This rwt file will work for any MPA study of a solid model we run in the future.

After you have experimented with result window definition files and templates, return to the main Creo Simulate window using *Close* on the Quick Access toolbar at the top.

Simulation Features in the Model Tree

Open up the model tree. Note that the simulation features (loads and constraints) are all listed. Expand the tree and you will see the members of the load set and constraint set created above. Additional columns can be added using *Settings*, as in Figure 33. If you right click on these entities and select *Edit Definition*, you can change any of these modeling entities directly out of the model tree. This is a very handy way to make changes to the simulation parameters. You can also, of course, select the model entity in the graphics window, then use the commands available in the RMB pop-up menu.

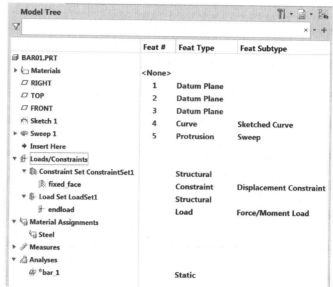

Figure 33 Model tree with simulation features

Keep these shortcuts in mind when you do some of the exercises at the end of this lesson.

Defining and Using Measures

Creo will automatically report the value of various key quantities, such as the total strain energy, maximum displacement magnitude or component, the maximum Von Mises stress, and others. These are reported in the **Study Status** window at the end of a run. Some of them are also used directly in the convergence analysis of the model (in SPA and MPA studies).

We saw previously that we could use various commands in the result windows (*Model Max*, *Dynamic Query*, etc.) to obtain information about the model. In addition, we might like to know things such as the stress or displacement at a particular point, or the largest value on a surface. It would be useful if we could obtain such data in the Study Status report. We can do this, and much more, using **Measures**. Even better, we can use measures explicitly in the analysis to control the p-loop convergence. We will see later that this is a very useful way to avoid some of the issues that arise using the built-in convergence controls.

For now, let's create a few measures for the bar model. Select

Home { Run } : Measures

then check the **Show Predefined Measures** box at the bottom. The predefined measures are the ones that are on the Study Status report. Let's add some measures of our own. As you go through the following, you should explore the many options available - it is important to know what is available[10].

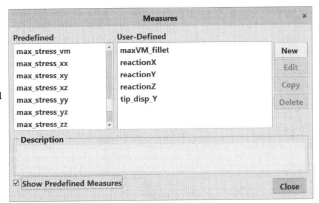

Figure 34 Lists of Predefined and User-Defined **Measures**

Select

New

Each measure has a unique name. For our first measure, fill in the dialog window as follows:

Name   *tip_disp_Y*
(Click *Details* and add a short description if you like)
Quantity   *Displacement*
Component   *Y*
(Note the default is the WCS)
Spatial Evaluation   *At Point*

Select the button under **Point(s)** and then on the model pick one of the top vertices on the free end of the bar, then select *OK* in the **Points Selection Window**. At the bottom of the Measure Definition window, it tells us what types of analysis this measure can be used for. Static analysis is listed. The completed dialog window is shown in Figure 35. Select *OK* (twice).

An icon appears at the end of the bar. If you have **Names** turned on in the **Simulation Display** window, you will also see the measure name.

We know that the maximum Von Mises stress on this model occurs on the inside curved surface of the bar. We are not sure exactly where that is. While still in the **Measures** window, select *New* again and enter the following

Name   *maxVM_fillet*
(Click *Details* and enter a description if you like)
Quantity   *Stress*
Component   *von Mises*
Spatial Evaluation   *Maximum*
                   Over Selected Geometry

---

[10] The author speaks from personal experience, since he spent many years doing a lengthy outboard calculation for a quantity that was already in the **Measures** list!

Now select the button under **Geometry** and pick on the surface on the inside of the corner of the bar. Middle click, then **OK**. An icon appears on the model (with measure name if turned on). This measure is set up to report the maximum stress value that occurs anywhere on this surface.

Figure 35 Measuring a displacement at a point

Figure 36 Measuring the stress on selected geometry

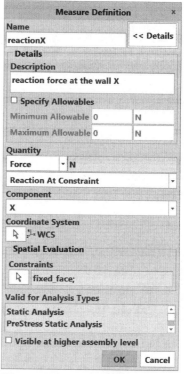

Figure 37 Measuring a constraint reaction force

Let's create some measures that will report the reaction forces on the constraint. Select *New* again and enter the data as follows:

> **Name**   *reactionX*
> (Select **Details** and enter description if desired)
> **Quantity**   *Force*
>   *Reaction At Constraint*
> **Component**   *X*
> (Note the default is the WCS)

In the **Spatial Evaluation** area, select the button under **Constraints** and pick on the constraint icon on the left end of the bar (or use the model tree entry). Middle click and then select **OK**.

Repeat the previous measure definition for the **Y** and **Z** components of the reaction. To make this a bit easier, check out the *Copy* command in the **Measures** window.

We now have a number of measures defined on the model. Use the **Simulation Display** window (check the **Common Settings** area) to turn on the display of the name tags. If you select each tag and hold down the RMB, you will see a number of commands (*Edit Definition*, *Suppress*, *Hide*, *Delete*, *Rename*) for dealing with measures. Use *Move Tag* to position the tag where convenient. The model might look something like the figure shown at the right. This is a great figure for documenting the model, say in a report!

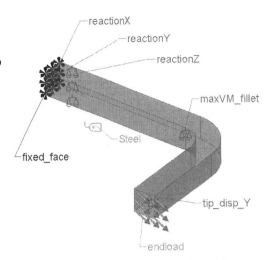

Figure 38 Model with a number of user-defined measures

Reporting Measures using Annotations

Unfortunately, measures cannot be calculated retroactively. We will have to rerun our most recent MPA design study. Do that now by selecting the analysis **bar_1** in the model tree, then in the RMB pop-up select *Start*. Open the **Summary** window by expanding to the right with the ">>" button. Near the end of the listing are all the predefined measures and at the bottom of that list are our user-defined measures:

maxVM_fillet	44 MPa
reactionX	-500 N
reactionY	250 N
reactionZ	(Something really small, essentially 0)
tip_disp_Y	-0.36 mm

Compare the stress and displacement values to the predefined measures. They are identical, but now we know exactly where they occur. Do the reaction values make sense? For this statically determinate problem, they should be equal and opposite to the applied forces. For a statically indeterminate problem, constraint reactions are difficult to determine analytically but are important bits of information since they express the forces transmitted to the surroundings by the component under study.

Now create (or open if you already have the *rwd* file) a fringe plot for the Von Mises stress. In the **Report** group, select

> *Measures*

A window opens showing all the predefined and user-defined measures in the model. Scroll down the list to find our measure **maxVM_fillet** and select it. A small gray circle appears on the model showing where this measure is evaluated. At the bottom of the **Measures** window select

Create Annotation

A box appears on the model showing the measure name and value, connected with a leader to the location mark.

Create another annotation for the tip displacement measure, then *Close* the **Measures** window.

Go to the **Format** ribbon. Make sure that *Show Annotations* is selected (also available in the Graphics toolbar). In the **Annotation** group, select *Edit*. Left click on one of the annotation boxes (it turns magenta), then middle click to accept. This opens the **Note** dialog window. To move the note, select the *Note Location* button and pick a new point for the leader elbow. Use the *Style* button to set text font and attributes (style and color), arrow style, note box style, background color, etc. See Figure 39.

Figure 39 A fringe plot with a couple of *Annotations* showing measure values

The location of each note is at a fixed point on the screen. If you spin, pan, or zoom the model, the attachment point on the model will move, but the note will not. So be careful!

In the **Annotation** group, the *New* button will allow you to create a note with your own text with the option of attaching it to the model with a leader. The *Mouse Sketch* tools allow you to put rudimentary markup on the result window.

Using Measures to Monitor and Control Convergence

Create a result window that shows how our measure **maxVM_fillet** behaves during the recent MPA analysis. Either copy one of the current convergence windows (convm, condef, or constr) or create a new one called [**conmeasVM**]. In the **Result Window Definition** dialog set the following

Display type	*Graph*
Quantity	*Measure*

Select the button under **Measure** to open the list of predefined and user-defined measures. Select our measure **maxVM_fillet** in the right panel, then *OK and Show*. In this case, it looks the same as the graph of the predefined measure **max_stress_vm**, since it is obviously (for this model!) measuring the same thing. The advantage of using our own measure is that we also know where this occurs and, as we will see in later lessons, our measured value is often more numerically reliable than the predefined measure

max_stress_vm which can be corrupted. Some of these issues will be discussed in the next lesson.

For now, the graph of our measure **maxVM_fillet** shows that after the 4th pass the value really doesn't change much - the model is very well behaved and we could even say this value has converged. This gives us the idea to monitor this measure in particular during the MPA, instead of the defaults, and terminate the run when it has converged. Let's try that.

Close all the result windows and open the **Analyses and Design Studies** window. Select the study **bar_1** and in the mini toolbar select *Edit Study*. The study should currently be set to MPA with 5% convergence and maximum order 9. At the bottom, note the default convergence measures. Check the radio box beside **Measures**, then select the button just below. This opens the **Measures** list where we can select the *maxVM_fillet* measure we created before. Actually, we can select more than one measure here and all would be required to converge before the solution stops.

Run the MPA again. As expected, the run stops on pass 4. The **Study Status** window reports the final value of **maxVM_fillet** as 44 MPa, essentially what we had before (and well within the 5% convergence tolerance we specified). So we have obtained essentially the same result with a model that executed in a fraction of the time of the original. On this simple model, that takes less than a second or two, that is not significant. But on a much more complex model, possibly requiring many minutes for each run (that may have to be repeated many times, say during an optimization), the time and resource savings can be substantial.

Custom measures are very handy tools for extracting specific information about the model. More importantly, they allow the analysis to converge and give best estimates on measures that are of greatest importance to us. You should spend some more time exploring the various options available for using measures and annotations for performing and reporting on your analyses. Some of these options will make the exercises at the end of the chapter a lot easier!

Exploring the FEA Mesh and AutoGEM

The FEA mesh created by the automatic mesh generator, AutoGEM, is shown in Figure 40. Let's see how to create this image. In the Graphics toolbar, select

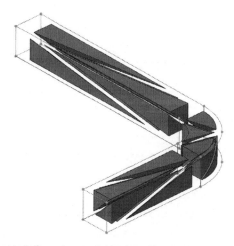

Simulation Display

In the window that appears select the tab

Loads/Constraints

and turn off all visibilities in the **Structure** area (loads, constraints, etc.) - use the button at the bottom to select all. Now use the **Mesh** tab, and make sure *Mesh Points* and *Solids* are checked, then select

Figure 40 Element mesh created by AutoGEM (default settings)

Shrink Elements 30%

In the **Modeling Entities** tab, turn off *Material Assignments* and *Measures*, then select *OK*.

Now we need to create the mesh. Go to the **Refine Model** ribbon and select

AutoGEM ▦ *> Create*

AutoGEM Summary and **Diagnostics** windows appear. Have a quick look, then close them. The display shows the mesh elements. People who have previously used other FEA packages may be alarmed at the geometry of these elements as follows:

- there are very few elements involved in the model
- some elements are very long and slender (high aspect ratio)
- element corners, even within the same element, can have very different angles
- transitions in element size through the mesh are quite abrupt (long, slender elements are adjacent to short, wide ones)

In an h-code based method, all these would be signs of a poorly constructed and possibly lethal mesh, and would raise serious concerns about the accuracy of the solution. In this section, we will explore different meshes for this model and compare the performance and results obtained. Hopefully, this will give you a bit more confidence in the results obtained with the default mesh, as coarse as it is.

Close the **AutoGEM** window. Do not save the mesh.

Before we proceed, let's review our results from the previous multi-pass analysis with the default convergence measures. Our main items of interest are the following:

number of tet elements	20 or 21
convergence (5%) on pass	7 or 8
maximum Von Mises stress	44 Mpa (approximately)
maximum displacement	0.55 mm

We are going to change the element mesh and compare results. Some of our control over the mesh is through the settings used in AutoGEM, the mesh generator. We will see other controls later. Go to the **Refine Model** ribbon and select the pull-down menu under *AutoGEM* then pick *Settings*. This brings up the dialog window shown in Figure 41. Select the **Limits** tab in the middle of the window. This produces the options shown in Figure 42.

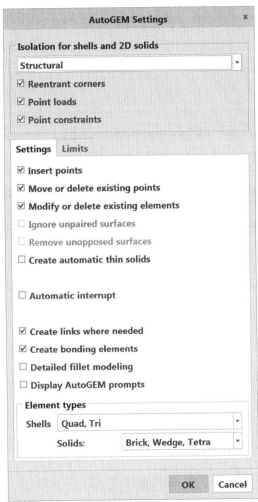

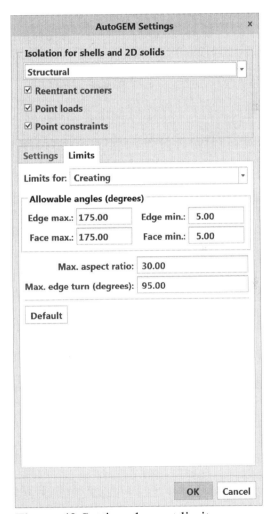

Figure 41 AutoGEM settings dialog window

Figure 42 Setting element limits

In the **Limits** area, there are basically three types of settings. The **Allowable Angles** are the angles between edges and faces of an element (see Figure 43). The **Aspect Ratio** is roughly the ratio of length to width of an element. The **Edge Turn** is the amount of arc that can be allowed on an edge (see Figure 44). As you can see, the default settings are pretty broad. The current settings account for the wide variation in geometry illustrated in the mesh in Figure 40.

If you tighten up on the allowed element limits, you should expect to see more elements in the model. Let's try that. Change the data in the **Limits** dialog to the values shown in Figure 42:

    Allowable Angles
        Edge   Max **150**   Min **30**
        Face   Max **150**   Min **30**
    Max Allowable Aspect Ratio   **4**
    Max Allowable Edge Turn   **45**

Then accept the dialog with *OK*. There will be a warning message about the aspect ratio. Note this in passing[11] and carry on.

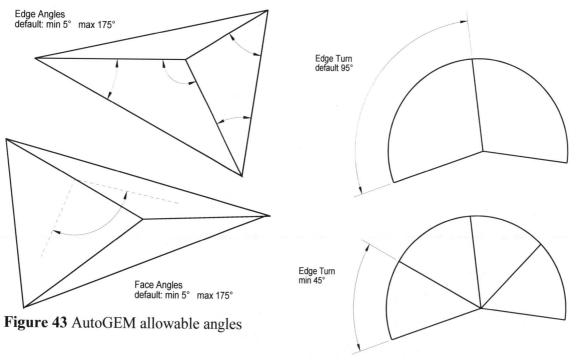

Figure 43 AutoGEM allowable angles

Figure 44 AutoGEM edge turn

We will go straight to the MPA (and skip the Quick Check). To make sure of settings, in the *Home* ribbon select

---

[11] Because of this warning, we very seldom use the aspect ratio control.

Analyses and Studies

and *Edit* the analysis **bar_1** to make sure our analysis is the same as before (Multi-Pass, 5% convergence on default measures, max order 9). Now select *Run Settings*. To make sure we use the new AutoGEM settings (and not the existing [20 element] mesh), check the boxes *Override Defaults* and *Create Elements during Run*. Accept the settings and press *Start* when ready. Delete the existing output files. This run will take a little longer than before - AutoGEM has to work a bit harder this time! Open the *Study Status* window.

The multi-pass run converges on pass 5 with a maximum edge order of 5 with the following results:

number of tet elements	180 (approximately)
maximum Von Mises stress	45 Mpa (approximately)
maximum displacement	0.56 mm

Check the **Log** tab to get even more information about the run. For example, a significant portion of the CPU seconds were used for mesh generation (see the elapsed time after pass 1) - it was about 70% of the total elapsed time.

---

* * * * IMPORTANT * * * *

Check the locations on the model for the maximum Von Mises stress and the maximum stress on the fillet. For the previous (default) mesh, these were identically the same. For the new mesh these values are almost the same but the locations are totally different. The lesson here is to be cautious about results that are reported while you are developing a model - make sure you understand what you are seeing. Be careful of assumptions you may be making without knowing it.

Use the AutoGEM command to display the mesh as we did before. You can retrieve the existing mesh by using (in the **Refine Model** ribbon)

> *AutoGEM*
> *File*
> *Copy Mesh from Study*

and select the MPA study (top-most directory) we just finished. This saves you some mesh-creation CPU time. The FEA mesh is shown in Figure 45. There are about 10 times as many elements in the model as before. As requested, there are no long, slender elements and no very small or very large edge angles.

Note that the run converged two passes sooner with a lower maximum edge order than the previous mesh. Generally speaking, the more elements in the mesh, the lower is the element order required for convergence. This is useful to know when you come across a model that will not converge even with the maximum edge order of 9 (NOTE: these cases are very rare unless the model contains singularities, which we will discuss later!). With lower orders, there are fewer equations per element, but there are more elements. In this case, the net result is that, although there are many more elements, the computation time (not including mesh generation) is (only) about five times what it was before. The relation between mesh size, edge order, and run time is very complicated!

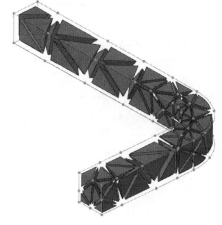

Figure 45 Mesh with custom AutoGEM settings

The Von Mises stress and maximum displacement values are within a few percent of the previous results. The displacement results differ by much less than the stress results. We could probably say that the results are equivalent. Remember that, above all, FEA results are an approximation to the real world. If we can be confident we are within 10% of the "real" value, consider the analysis a success. Convergence plots for the run are shown in Figure 46. If you previously created a results window template, this can be easily retrieved by selecting (in the **Home** ribbon)

Results > Open from Template

Specify the current design study (**bar_1**), then the template **MPA_standard.rwt** that we created earlier. Finally, select *OK and Show*. All the windows defined in the template are available for display. You can choose a subset of these using the *Show* and *Hide* buttons on the **View** ribbon (see Figure 18).

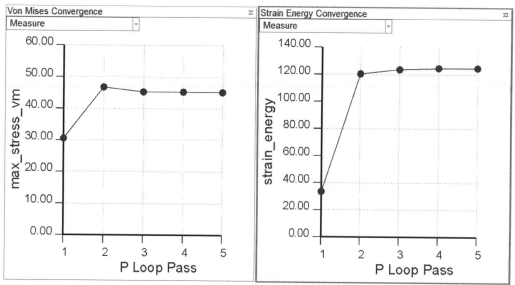

Figure 46 Convergence of Von Mises stress (left) and total strain energy (right) using the custom mesh

Observe in Figure 46 what would happen if this problem was solved using an FEA model having only first order (linear) elements. Many (naive) users might be satisfied with the mesh shown in Figure 45 and probably with these results too. However, without doing a convergence check, they would be stuck with the results obtained on pass 1, not realizing they are in error by almost 50% on the maximum Von Mises stress.

There are clearly two lessons to be learned here. The first is a general observation: FEA results are useless unless accompanied by an examination (and demonstration) of convergence of the solution.

The second lesson is that, at least in this problem, the mesh density in Creo did not have a large effect on the results in the final solution, although it did affect the solution time. The extreme variations in mesh geometry (aspect ratio, skewness, mesh transition, etc) observed in the default mesh do not cause problems in the solution, as would happen in an h-code solution. In the future, you can probably be comfortable using the default settings for AutoGEM. There may be occasions when you will have a good reason to modify some of these settings, such as when it is difficult or impossible to obtain convergence with the maximum edge order. An example is the allowed edge turn. In a part with filleted corners in a high stress region, you may want to reduce the edge turn to have more elements along the fillet. The same applies if you have holes in a part in a critical region. Before you change AutoGEM settings (that apply everywhere on the model), have a look at the mesh generated by default. There are routines in AutoGEM that will automatically use a denser mesh in such areas. See, for example, the AutoGEM option "Detailed Fillet Modeling." There are other mesh controls that we will discover as we go through the lessons that produce more specific mesh refinement than the global settings we used here.

Summary

In this lesson, we have performed a complete static stress analysis of a simple part, using solid elements. We have seen all the major steps involved in carrying out the solution and most of the important Creo command menus and dialog windows. To recap, these major steps were the following:

1. Create the geometry in Creo Parametric, then launch Creo Simulate.
2. Select the model type.
3. Complete the model using
   a. constraints
   b. loads
   c. material properties.
4. Define an analysis and specify mesh and convergence properties.
5. Run the analysis.
6. Define and review the desired result windows.

These are basically the same steps you would go through if you were to launch the Creo **Process Guide**. Go to the **Tools** ribbon and in the **Process Guide** group, select *Run*

Session. Pick the process template

StaticAnalysisNative.xml > OK

Although this is probably not a good way to learn how to do an analysis, it would be a useful tool as a reminder or if there were some "tricky" bits to setting up the model. These process guides can be customized or new ones written from scratch to support some user's special modeling needs.

There are many options to explore in the sequence given above, and many variations of the result windows which you can investigate on your own (perhaps while doing the exercises!).

We observed the effects of changing the global settings for the automatic mesh generator. Since the behavior and accuracy of the solution will depend on the mesh geometry (especially for non-converged solutions), we will be examining a number of ways to customize or fine tune the mesh to get better results.

In the next lesson, we will look at the different types of design studies (standard, sensitivity and optimization), as well as looking at multiple load cases and superposition of solutions. Some important issues in defining loads and constraints will be addressed.

Questions for Review

1.   What form of geometric primitive does AutoGEM use to create solid elements?
2.   Are there any restrictions on the curvature of the edges of any solid element?
3.   Can you get AutoGEM to create brick (8-node) or wedge (6-node) elements?
4.   In the AutoGEM settings area, what is meant by edge angle and face angle? What are the default values for their minimum and maximum values?
5.   See if you can find out how the aspect ratio of a solid element is determined.
6.   How do you tell Creo what the part being studied is made of?
7.   Sketch the constraint symbol and identify the six boxes.
8.   The first analysis usually performed is simply to determine if there are any errors in the model. This is called a _____. How is this set up?
9.   What is the purpose of a multi-pass adaptive analysis?
10. How do you set up a window to view the results of an analysis?
11. How can you find the location of the maximum stress in the model?
12. What is the command for a "live" numerical readout of the stress at the cursor location?
13. What is the difference between a cutting surface and a capping surface?
14. What is the main benefit of seeing an animated deformation?
15. What is meant by "convergence"? How is this set up? What are two ways of examining the convergence behavior?
16. What is the meaning of the following buttons:

     a) ▨   b) ▨   c) ⊢   d) ▨   e) ▨

17. What is our first remedy to a static analysis that will not converge in the allowed maximum number of passes?
18. What material properties must be defined for a static stress analysis?
19. Consider a long circular bar loaded in pure tension. The axial stress depends only on the geometry of the bar (that is, the cross section area). Why, in FEA, do we have to specify what the bar is made of? Does this make a difference? What effect, in general, does the material definition have on the solution?
20. Find out what result windows are produced by the system default templates.
21. What is a *Measure*, and how do you create it?
22. What are the main benefits of user-defined measures?
23. How can you use a measure to determine when an analysis has converged?
24. How can you determine the support reactions in a statically indeterminate problem?

Exercises

The objectives of these exercises are:

    a) to get comfortable with the basic tools, procedures, and utilities for performing static stress analysis;

    b) to discover some of the limitations of both analytical models and FEA models;

    c) to begin to critically examine results obtained and relate them to fundamental theory and physical expectations.

Read through each exercise before starting - hints are sometimes buried in the text.

---

1. A deceptively simple model with surprising "diabolical" behavior for the unwary!

Consider a 200mm long cantilevered beam with rectangular cross section (width 20mm, height 40mm). A vertical load of 40 N is applied downward at the free end and the left face is fully fixed. The material is steel. This is a very simple model for which there is an analytical expression (from simple beam theory) for the deformation and stresses. Set this up so that the X axis is on the neutral axis with the left end at x = 0.

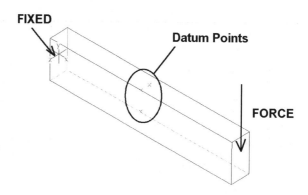

Do not create the Datum Points just yet - we are going to use them later.

We are interested in the stress values (probably normal stress σ_{xx} will be most useful, although the shear stress τ_{xy} may also be of interest), strains and displacements (consider all directions). In the following, observe the convergence behavior and any other special occurrences in the solutions. Using this simple model geometry, we will explore several aspects of interest (convergence settings, mesh refinement, custom measures, material properties). Find out what happens in the following situations:

(a) **Base line model**: fixed surface at left end, uniform load on right end, no extra datum points, default mesh. Use MPA with 5% convergence on default measures for this run. Compare these results to the analytical solution where possible[12]. Is the analytical solution "exact"? Why? (Hint: what assumptions and simplifications are required to produce the analytical model? How do these differ from those used to create the FEA model?)

(b) What happens if you try to use a much smaller convergence (like 1%)? Why? (Hint:

---

[12] A very useful output graph is to plot a **Graph** of a **Stress** component value along a specified **Curve**. In the case of the beam, plot a graph of σ_{xx} along one of the top or bottom edges.

show an animation of the deformation using a fringe plot of the displacement component in the Z direction.) Comment on the notion of "pushing" the solution[13].

(c) Implement some mesh refinement using the settings for AutoGEM. For example, change the minimum and maximum edge and face angles to $30°$ and $150°$, respectively. (Avoid the aspect ratio control, since it sometimes behaves very strangely; edge turn will have no effect on this model). Compare the convergence behavior of the 5% and 1% MPA for this new mesh.

(d) Now create the datum point(s) at the mid-span location. Create some measures (like displacement, Von Mises stress, τ_{xy}, and/or σ_{xx}) at the datum points. Do these values agree with the analytical solution? What is their convergence behavior? Perform MPA convergence on one or more of those measures (which ones make sense? Hint: do not try to converge on a value of 0, for example σ_{xx} measured on the neutral axis). Is this a good idea? Can you push this convergence to 1%? Does the tighter convergence tolerance produce a "better" solution?

(e) Material properties are often taken for granted. To explore the effect of material properties, go back to the most accurate, best-behaved previous model and independently change the values of Young's modulus and Poisson's ratio for the assigned material. (Hint: select *Materials* in the **Home** ribbon, and *Edit* the material definition). Make a note of what you expect to happen to the stress state and deformation. First, change E by factors of 0.5 and 2.0. Then, with the original E, try values of Poisson's ratio of 0.2, 0.3, and 0.4 (can you get a solution with 0.5? Why?). What effects do you note? How much change in these properties is required to cause a significant change in the results?

---

2.   Exploring the limits on a model

An "interesting" situation occurs if you try to perform an MPA analysis on our L-shaped bar using the modified mesh and a convergence criterion of 1%, maximum order 9. Try this and plot the usual convergence plots. Where does the maximum computed stress occur? Can you explain what is happening in the model? (Hint: consider the effect of Poisson's ratio by plotting an animation of the transverse displacement.) Can you avoid this problem? If so, how?

---

3.   A fairly straight forward solid model

Compute the maximum Von Mises stress and deflection in the steel hook shown in the figures below. Model the hook as a sweep cantilevered out from the fixed top plate. (Do you need to model the top plate?) Dimensions are in inches. Examine the convergence behavior of the model. Can you find an analytical solution for this problem, using expressions for curved beams?

---

[13] "Pushing" refers to trying to get a more accurate solution than, perhaps, the model or software is capable of providing.

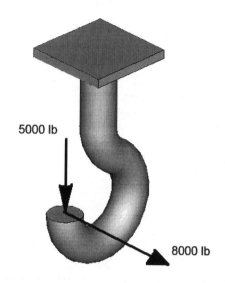

5000 lb

8000 lb

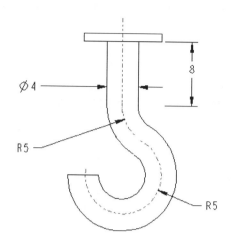

⌀4

R5

R5

8

---

4.   A typical FEA problem

Compute the maximum Von Mises stress and deflections in the directions of the loads on the end of the steel connecting rod. The loads are applied at the smaller end. Assume that the inner surface at the larger end is fixed. Examine the convergence behavior. Dimensions are mm. What happens if you put a R2mm round where the bar meets the two cylindrical ends? This is a case where too much defeaturing can cause issues with the FEA model.

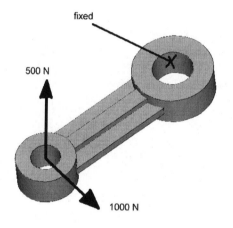

fixed

500 N

1000 N

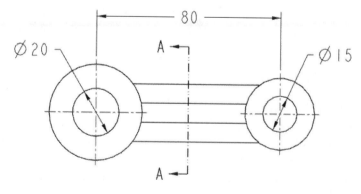

⌀20

⌀15

80

A

A

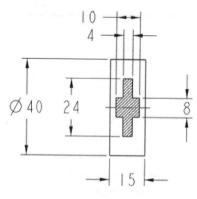

10

4

⌀40   24

8

15

SECTION   A-A

5. Exploring options for applying loads and mesh generation

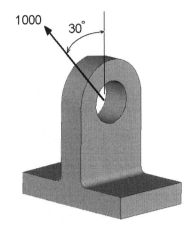

Examine the behavior of the aluminum bracket model at the right. Dimensions are in the figure below with units in inches (force in pounds). Assume the vertical faces at each end of the base plate are fully fixed. Look at the stresses, deformation, and convergence. Experiment with the following variations:

a. Model the geometry exactly as shown in the figures. Use all default settings for mesh generation and the static analysis convergence. Apply a *Uniform | Total* force to the inside surface of the hole using the magnitude and direction given at the right.
(HINT: when defining the force load, select
   Dir Vector & Mag
in the pull-down list. Enter the components of the direction vector as **cos(30)** in the Y direction and **sin(30)** in the Z direction, with a magnitude of **1000**.)

b. Remove the previous load and try out the *Bearing* option in the **Loads** menu. This distributes the total load normal to the bearing surface of the hole. Use *Preview* to observe the way the load is distributed. What effect does this have on the stress in the base plate? Around the hole?

c. Examine the effects of the AutoGEM setting **Max Edge Turn** (set it to 30) on the mesh and the results.

d. Suppress the fillet and measure the maximum stress and displacement in the model. Plot the convergence graphs (especially the Von Mises stress graph). Comments?

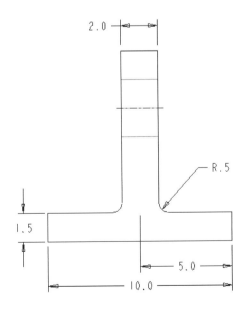

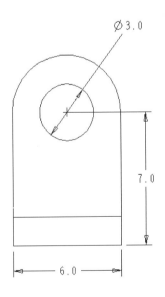

6. The documentation for Creo Simulate recommends that **Single Pass Adaptive (SPA)** analysis be the one normally used. Repeat some of the analyses of exercises 1 through 5 above using the SPA convergence analysis type. This should only require a minor modification in the run settings. Don't forget to provide a new name for the design study so you don't overwrite the previous results - then you can show results of both studies simultaneously. Many of the same result windows will still apply. Convergence graphs will obviously not be possible for SPA convergence. What are the main differences in the behavior of the model or the results obtained? Are these significant? An interesting result window to show is a fringe plot of the element P-levels. How successfully does SPA convergence analysis deal with some of the diabolical behaviors demonstrated in the previous models?

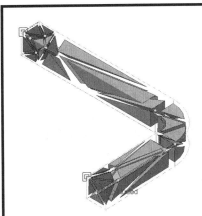

Chapter 4 :

Solid Models - Part 2

Design Studies, Optimization,
AutoGEM Controls, Superposition

Overview of this Lesson

In the previous lesson, we performed a simple static stress analysis of a part with given loads and constraints. In this lesson, we will first look at the three types of design studies: *Standard Studies, Sensitivity Studies* and *Optimization.* The purpose of these design studies is to simplify and, in the latter two types, automate some of the repetitive work involved in design. Each of these studies involves specifying one or more design variables (typically dimensions) that control the geometry of the part. The design variables have specified values, ranges, or limits. The variables can be used by any type of study as follows:

Standard - the model is evaluated for a single set of values of the design variables. If a single specific geometry of the model is of interest, this would be the type of study to use.

Sensitivity - the model is evaluated for a specified range of values of the design variables. This would be done to find out how much a variable affects the results of interest (like the maximum stress or deflection).

Optimization - this is set up to find the values of the design variables (within specified ranges) that will achieve some design goal, like minimizing the part mass, while not exceeding some design constraint, like the maximum allowed stress.

We will also look at some important considerations when applying loads and constraints. These are related to the use of p-elements in the model and the effect on the convergence analysis. Specifically, we'll examine what bad things happen if we specify point and edge loads or constraints in a solid model and how these problems can be avoided by excluding elements from the convergence analysis.

Finally, we will set up and use multiple load sets in order to generalize a solution. This relies on the principle of superposition of solutions. As usual, there are some Questions for Review and some Exercises at the end of the chapter.

Standard Design Studies

Suppose you want to find out how a particular dimension (size or location) or model property will affect the results of an analysis. If there is a particular value of a dimension, for example, that you want to investigate, you could return to Creo Parametric, edit the dimension, regenerate the part, come back into Creo Simulate, and run the static analysis as we did in the previous chapter. This is obviously a lot of steps. A *Standard Study* allows you to do all this within Creo Simulate.

The general procedure is to set up the model as usual - create the geometry, generate the elements, specify loads, constraints and material properties, and create an analysis definition (like static stress, modal, etc.). You then pick the dimensions you want to specify as *design variables*. In a Standard study, you can then set the value of each design variable and run the specified analysis without returning to Creo Parametric to regenerate the model.

Let's see how all this works. Launch Creo and set the working directory to **CHAP3** (if you have been following instructions to the letter). Bring in the part **bar01** that we used in the previous lesson.

To make it a bit easier to follow the procedure, it is useful to change the symbolic names of the dimensions in the model. Select the sweep feature in the model tree then *Edit Dimension* in the mini toolbar. Pick the dimension for the cross section (**25**). Then in the **Dimension** ribbon at the top, change the symbolic name to **width**. Accept this and then select the sweep trajectory in the model tree. Change the dimension symbol for the arc on the corner of the trajectory to **bend_radius**. Use *Switch Symbols* (in the **Tools** ribbon) to verify the change. These modified dimension symbols are shown in Figure 1.

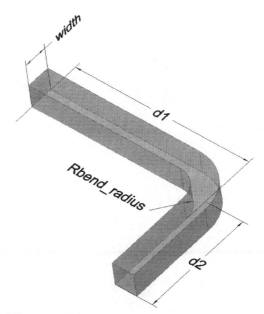

Now we can transfer into Creo Simulate:

Figure 1 Dimension symbols renamed

Applications > Simulate

If you have been playing with the mesh, go into the *AutoGEM > Settings* menu (in the **Refine Model** ribbon) and under the *Limits* tab change the settings to the defaults. Check that the load, constraint, and material are defined.

If you recall in the previous chapter, our multi-pass analysis **bar_1** converged on essentially the 3rd or 4th pass, although we let it run a bit longer. So, let's modify the convergence analysis so that it stops then. If your analysis has been deleted, create a new one called **bar_1** using

Analysis and Studies > File | New Static

For **bar_1**, make sure load and constraint sets are selected. Set a **Multi-Pass Adaptive** convergence to 10% on the default measures. Set the maximum polynomial order to **5** and accept the dialog with a middle click. We are being a bit less stringent in our requirements here in order to decrease the execution time. We know from the results of the previous chapter that this will not significantly affect the final results. Accept the definition. You should come back later to see what happens with a refined mesh and using your own measures for convergence as we did in the previous lesson.

Use *Info > Check Model* to see if there are any gross errors. You might like to check the setting to *Create Elements During Run*. Run the new analysis **bar_1** just to make sure it is doing what we want. The run should stop on pass 4. The results should be more or less the same as we obtained previously (except that the run will take a lot less time): maximum Von Mises stress around 44 MPa, maximum displacement magnitude 0.55mm.

Creating a Design Variable

In the **Home** ribbon, select *Analyses and Studies*. The static analysis **bar_1** should be listed there. In this window, select

File > New Standard Design Study

This opens the dialog window shown in Figure 2. Enter a new name of the study (**"width50"**) and a suitable description. Select the top button to the right of the **Variables** pane - this is the *Select Dimension* button. Click on the part to reveal the width dimension (currently 25). Click on it to add it to the list of design variables displayed in the pane. This shows the new dimension name we created previously and its current value. In the column under **Setting**, change the value to *50*. In the **Analyses** pane above, select the static analysis **bar_1**. The other option here, as the name

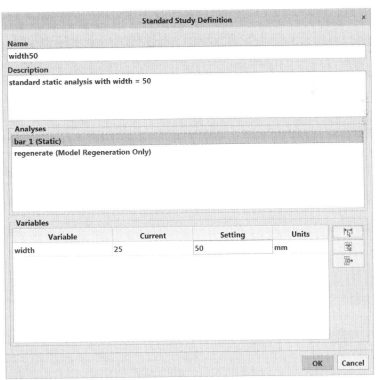

Figure 2 Setting up a Standard design study

implies, will just cause model regeneration with no analysis - this might be useful if you have a complicated model and want to make sure the specified variable setting(s) work

properly. See Figure 2 for the completed dialog window. Accept the study definition with
OK or middle click.

Back in the design study definitions window, select our new study **width50**, and then the
Run button in the top toolbar. This run will take a bit longer than the simple static
analysis since Creo Parametric must be launched in the background and the part
geometry regenerated with the new width value. Open the **Run Status** window to watch
the proceedings. AutoGEM creates something like 21 elements and the run stops on pass
5. When the run is completed, create the normal result windows (if you created the
template in the previous lesson, you can very easily do this with **Results > Open from
Template**). These will look something like the results shown in Figure 3. After
examining these, close out the results windows and return to the *Analyses and Design
Studies* dialog window.

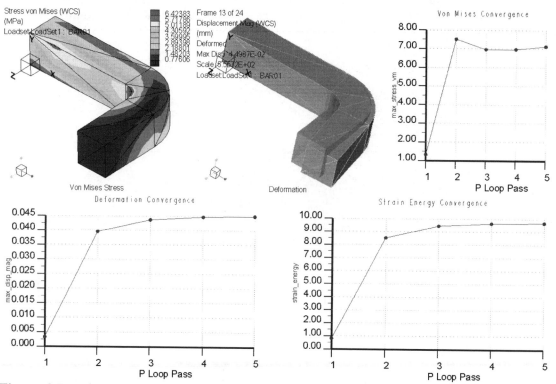

Figure 3 Results windows for **width50** Standard Design Study

You can now edit the Standard design study to set whatever value of the width variable
you want. This is much more convenient than editing the model and regenerating back in
Parametric, with the added advantage that the model geometry stored in the part file
stored on disk is not actually altered. You should note, however, that the part geometry
used for the **width50** design study is saved in the **width50** subdirectory along with all the
run results. If you do not want to overwrite that data, change the name of the standard
design study for each new set of parameters (like **width10**, **width20**, etc.).

Sensitivity Design Studies

In a Standard study, you can pick a single value for each defined design variable. Suppose you want to know what will happen to the model when a part dimension varies over a specified range - you want to assess the sensitivity of the model to changes in this parameter. You could do this by manually editing the model (geometry or properties) and performing the Standard study many times. The purpose of a Sensitivity Study is to automate this task.

In this sensitivity study we are interested in finding out the effect of the bar's cross section dimension (**width**) on the maximum Von Mises stress and total mass of the part. Rather than giving a single value for the section width, we will specify the range over which it should change. When the study is run Creo will automatically increment the specified design variable(s), manage the model geometry (i.e. regenerate when needed), and run the designated analysis on the model for each new configuration. You can then set up a results window to show the variation in some measure (like maximum Von Mises stress in the model) as a function of a designated design variable.

In the home ribbon, select the *Analyses and Studies* button then

File > New Sensitivity Design Study

This opens the dialog window shown in Figure 4. Call the new design study *[bar_1s]* ("s" for sensitivity). Enter a short description. The default type is a **Global Sensitivity**. The analysis to be used is **bar_1**, as defined previously. To the right of the **Variables** pane, pick the *Select Dimension* button, and add the section width dimension to the variable list. The current value is 25. Set the **Start** value to *20* and the **End** value to *35*. At the bottom of the window, set the number of steps in the study to **6** (so that the increment between values of **width** is 15.0/6 = 2.5mm). See Figure 4 for the completed window. Don't close this window just yet.

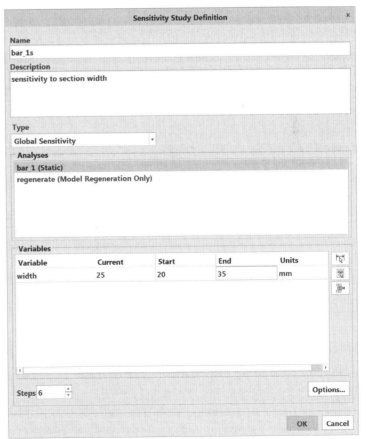

Figure 4 Setting up a Sensitivity Study

Performing a Shape Review

Although we used the word "automatic" in the description of the process, some subtle problems can sometimes arise in running a sensitivity study due to the changing geometry. Chief among these involves the element mesh in the model. As much as possible, Creo tries to use the same mesh throughout the study (to save time). Since, as we have seen, the solution is quite tolerant of large variations in the mesh geometry, this is usually not a problem. However, you must be cautious about the possibility that certain combinations of design variables may result in impossible meshes. If the geometric variation is large, there may be trouble. For example, if the design variables are the diameters of two holes in a plate, then the locations of the holes and values of the diameters must not allow the holes to intersect. Creo offers tools to check for these types of problems, and the solutions are often easy to obtain. A related problem, not specifically of Creo's making, is whether the model can actually regenerate with desired design variable values. For example, we will shortly be picking the bend radius dimension as another design variable. It would not make sense, and we should expect trouble, if we requested a value for this radius larger than the length of the bar. This is an issue of robustness of the model itself.

Creo offers a tool to help explore the design variables. In the **Sensitivity Study Definition** window, select the button *Options* at the lower right. A couple of the options are grayed out (since they apply only to optimization). You can leave the **Repeat P-loop Convergence** option check box blank. What this option does (if checked) is force a complete convergence analysis (MPA or SPA) at each new value of the design parameter. This might be necessary if you are concerned that changes in the geometry might cause problems with convergence. If unchecked, the design study will perform a multi-pass analysis (as defined in **bar_1**) for the first value of the design parameter and, since the mesh doesn't change, use these polynomial orders immediately for each subsequent value of the parameter. This will obviously be quicker. A second option **Remesh After Each Shape Update** is pretty self-explanatory.

To make sure the model will "survive" changes in the design variables over the specified range (and that we haven't accidentally given an incorrect value, by slipping a decimal for example), select the button *Shape Animate the Model*. All design variables will be set to their Start value and the model regenerated. It is now displayed on the screen with the width set to 20. We can now step the model forward through the desired range of design variables. Just middle click, or press Enter to go to the next value. Eventually, you will reach the end value and you are asked if you want to restore the model to its original shape. Select *Yes*.

If the model cannot be regenerated with some (or set of) design variables, the sequence of steps will be terminated. An error message will indicate the failure to regenerate and the model is returned to the initial values.

If you have several design variables defined in the model, they will all be incremented simultaneously over each one's specified range. To see the effect of a single variable, set the start and end values of the others to be the same.

Performing a shape review is important for sensitivity studies. As we shall see, this is even more critical for optimization, where the number of design variables is usually greater and they can interact in unexpected ways. Curiously, the *Shape Animate* function does not appear in the **Optimization Design Study Definition** window we will encounter a bit later. In order to perform the shape review, you will have to come back here to the sensitivity study definition window. *Close* the Options window.

With the sensitivity study definition complete (don't forget to select the analysis *bar_1 (Static)* as in Figure 4), select *OK*.

Running the Sensitivity Study

Back in the **Analyses and Design Studies** window, there are now three entries in the run list - our static analysis **bar_1** , the standard study **width50**, and the sensitivity design study **bar_1s**. Select the latter, and then *Start Run*. Open the *Study Status* window and watch the proceedings. Observe the convergence analysis is performed only on the first step. For the second and subsequent steps, it goes immediately to the highest order. While it is running, read on....

What is happening to the mesh as the geometry changes? In Creo language, we say that the mesh is *associated* with the geometry. This means that the mesh is attached to geometric curves and points, and changes shape as the model changes shape each time it is regenerated with a new parameter value. Creo does not have to recreate the mesh with AutoGEM for each new value. This is what gives Creo so much flexibility. As we saw in the previous chapter, the p-code mesh is very forgiving of large changes in mesh structure. This will be even more important in the optimization we will perform a bit later.

The run will complete in a few seconds. You will note that the total elapsed time is many times larger than the actual CPU time. The difference is related to the overhead in loading Creo Parametric afresh and regenerating the model for each new parameter value. Close the Run Status listing and select *Close* in the **Analyses and Design Studies** window.

Displaying the Sensitivity Results

Select the *Results* button (in the **Run** group of the **Home** ribbon) to create a couple of result windows. Select the *Open* button. The **Design Study** should be **bar_1s**. Call the window *[vm]* and fill in the rest of the dialog window as shown in Figure 5. Creo has already guessed that you want to plot a **Graph**. The Quantity is a **Measure**; use the button to pick **max_stress_vm**. Note that you can also choose from user-defined measures. The Graph Abscissa is automatically a Design Variable (Creo has figured out that **bar_1s** is not a standard analysis), and the **width** parameter is automatically chosen (there is only one in the model). Select *OK and Show* the window.

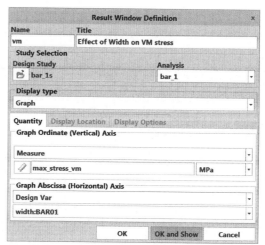

Figure 5 Defining a result window for the sensitivity study

Copy this result window definition to another result window called *[mass]*. Set it up to plot the measure **total_mass** against the **width** parameter. Don't forget to change the title.

Create a third window *[dispY]* that will plot the maximum deflection in the Y direction (**max_disp_y**).

Note that you could also select a measure for the horizontal axis. You could, for example, plot the maximum deflection or stress as a function of total mass. This might be of interest in problems involving questions of weight vs stiffness of a structure.

Showing the Result Windows

You can select the three window definitions in the view list and show them all at once. The graphs are shown in Figures 6.

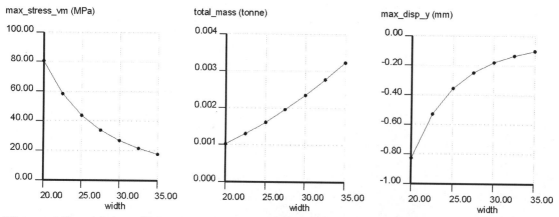

Figure 6 Sensitivity of Max Stress VM, Total Mass, and Max Displacement to width

As expected, as the width of the section increases, the Von Mises stress goes down, the mass goes up, and the Y displacement gets smaller (notice that the displacement graph is plotting negative values, with positive Y upwards in the model).

A sensitivity study would allow us to select an appropriate dimension to obtain a desired stress level, for example. If several variables are being investigated, we can see which of them has the largest effect on the stress. To do this, separate studies must be done with each variable. It is possible to have several variables included in a single study, but they are all changed simultaneously and the effect of an individual variable may be hard to determine.

When you have one of these graphs displayed (or window activated), select *File > Save As > Save a Backup*. In the Type dropdown list at the lower right, select *Graph Tabular Report*. This will let you write the graph data to a text file, with extension ".grt". Note you can also export to a spreadsheet format.

We are finished with this sensitivity study. Before leaving the topic, recall that we performed a **Global Sensitivity**, which involves variables ranging over a specified range of values. A **Local Sensitivity** study is used to assess which variables have the greatest effect on a measure at the *current* variable values. This

Derivative of ...	(With respect to) width	(With respect to) bend_radius
max_vm	- 4.7	-0.46
max_disp_Y	+0.054	+0.00047
total_mass	+0.000133	- 0.0000021

varies the selected variables(s) by a small amount around the current value, telling you which variables will have a larger effect on the particular measure of interest. This is essentially measuring the derivative of the measure with respect to the design variable. For our L-shaped bar, a local sensitivity study with parameters **width** and **bend_radius** produces the results in the table above[1]. It is easy to see that the width variable has a much larger effect on all listed measures (stress, mass, displacement) than the bend radius. Furthermore, the sign of the entry tells you whether variables will increase or decrease the measure if the model is taken away from its current design point.

Return to the main Creo Simulate window.

---

[1] Values in this table were obtained using 5% convergence and max order 7 in the static analysis.

Optimization Design Studies

In this exercise, we want to determine the values of the **width** and **bend_radius** variables that will result in the minimum weight bar, without exceeding a specified value of stress. From the local sensitivity study we know that:

>   increasing **width** results in lower maximum stress but increased mass

and

>   increasing **bend_radius** results in lower maximum stress and slightly lower mass

So we have an idea what needs to be done with these parameters. The question is how much to change them to produce an optimal solution. This calls for an optimization study. These are the most computationally intensive studies. Before proceeding with an optimization, you will usually perform a couple of standard analyses, and maybe even a sensitivity study[2], to determine the behavior of the solution. For example, you should have some idea as to the convergence behavior of the model. It also helps if you have a pretty good idea of where the optimum solution lies so that you can reduce the size of the optimization search space. This does not affect the optimization algorithm as much as it does guarantee the ability to regenerate the model for any point in the search space.

Creating Design Variables

We need to set up another design variable for the radius of the bend in the trajectory. In the **Home** ribbon select

>   *Analyses and Studies*
>   *File > New Optimization Design Study*

The optimization study definition window opens (Figure 7). Change the name to **bar_1opt** and enter a description.

The default study is an **Optimization** and the default goal is to minimize the total mass[3]. Notice that many other possible goals could be set up (such as maximizing an inertia, for example).

In the **Design Limits** area, to limit the maximum stress in the model, select the button on the right to add a row, and in the **Measures** window select **max_stress_vm**. This is

---

[2] There is no point trying to optimize on a variable that has very little effect on the model.

[3] A **Feasibility** study will find a solution that satisfies the constraints but without searching for a goal like minimum mass. In this case, from the current values it will find width = 27.0 mm, bend_radius = 26.3 mm, mass = 1.85 kg. Compare these to the optimized solution found later.

inserted in the table. Keep the default " < " sign and enter the limiting value *35*. Notice the default units. If you recall the maximum stress currently produced in the model, you will note that the model currently violates this limit. This observation is important - unlike many optimization techniques, we do not have to start the optimization from a feasible solution (i.e. one that does not violate a design constraint).

Beside the **Variables** pane, pick the *Select Dimension* button at the right to add the section **width** dimension. Set the minimum value to **20** and the maximum value to **35**. Pick the *Select Dimension* button again, and add the dimension **bend_radius** to the variables list by selecting the curve that was used for the sweep trajectory (easiest to pick in the model tree). Use a minimum value of **20** and a maximum value of **60**. This will give a fair bit of search latitude.

Select the analysis **bar_1** for the stress evaluation. This will be the default if there are no other analyses available.

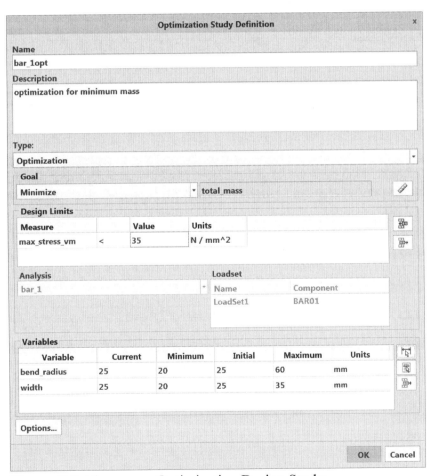

Figure 7 Setting up an Optimization Design Study

Check out the ***Options*** button in the study definition. Leave the optimization algorithm set to **Automatic**, and the convergence and iteration limit set to default values. The optimization routine will typically converge in under 10 iterations. Given the other assumptions and approximation made in the FEA model, you might also consider relaxing the default requirement for a 1% convergence on the optimization.

Important Considerations for the Search Space

It is curious that the shape review function does not appear in the Design Study Options window for optimization. It is very important to make sure that your part/model will regenerate for all combinations of values of the design variables within the search space (or at least those most likely to be encountered). Unlike for a sensitivity analysis, we really don't know beforehand which points within the search space will be visited during optimization.   For our simple bar, this should not be a problem due to the way the model was set up. You will recall that it was important for the swept section of the bar to appear on the outside of the arc at the bend. Imagine the difficulty you could get into if the section was on the inside of the arc and we tried to evaluate for a section width larger than the bend radius. The part regeneration would fail and our optimization procedure would terminate unsuccessfully. That would not be considered to be a robust model. The more design variables you use, the more complicated this can get. This is yet another issue that the model designer must keep in mind!

It is important, therefore, to do a shape review to ensure that the part is "regenerate-able" over a wide range of geometries, otherwise the optimization routine will stop in midstream. In order to do this, you could create a global sensitivity study definition, specify the desired design variables, and use the ***Shape Animate*** function available there as an option. Do this now to see how it works. The disadvantage of a global sensitivity shape review is that all parameters are incremented simultaneously and proportionately from the minimum to the maximum values. It would not show a shape where parameter A was at a minimum while parameter B was at a maximum. For these geometries, you could use the **Regenerate Only** option in a Standard design study.

Another possible issue during optimization concerns the mesh. One thing that is not available in integrated mode during a shape review is a look at the elements in the mesh. With very wide variations in parameters, and since by default the mesh is not recreated for each new geometry but stays associated as discussed above, it is possible that some combinations of parameters may result in geometries that have illegal meshes. If this is a possibility you can force a remesh of the model with each variable update, which takes additional time of course.

The way around the problems described above is to not use extremely wide ranges in the variables (that is, you should know roughly where the optimum solution exists) or use a Standard Design Study and a Quick Check analysis for the extreme geometries. At the end of the review, restore the model to its original shape.

Close the options window and the optimization definition window.

Running the Optimization Design Study

You can now **Run** the design study, **bar_1opt**. This could take a couple of minutes depending on your system. You can follow the process in the **Study Status** window. While this is going on, read the next section. The vast majority of the elapsed time for this model is spent loading and unloading Creo Parametric in the background so that the geometry can be modified. (Look for the trail files in your working directory!) In a more complex FEA problem, the fraction of time spent on these parameter updates would be considerably less as more time would be spent crunching each static analysis.

What Happens During Optimization?

You may be familiar with simple numerical optimization algorithms such as the method of steepest descent. The algorithm in Creo is considerably more complex than this, although the basic idea is the same. The algorithm is considerably more efficient than simple steepest descent, and also must contend with the limits (known as *constraints* in optimization theory) in the search space. Creo evaluates the current design and tries to decide in what direction to move in the search space (and how far) in order to either remove a constraint violation (like exceeding the allowed stress) or improve on the goal (in our case to reduce the mass).

According to the documentation, you can select from two optimization algorithms: the sequential quadratic programming (SQP) algorithm and the gradient projection (GDP) algorithm. The default is the SQP, which is generally faster for problems with multiple design variables. If the initial design point is feasible (that is, no constraints are violated), the algorithm moves the design point in a direction to better satisfy the goal until/unless a constraint boundary is met in the search space. Then it moves in a direction tangent to the constraint surface, all the while seeking out the minimum value of the objective function. If the initial design point is infeasible (constraints are violated), then one or more correction steps are taken to reach the (nearest?) constraint boundary. Thus, if the first design is infeasible, the design at the end of these first iteration steps is not guaranteed to be feasible. The GDP has the advantage that, if started with a feasible design, it tends to produce a series of intermediate designs that are always feasible, even if it is unable to locate the global optimum design (either due to the objective function or limits set by you). In contrast, the SQP algorithm does not guarantee that intermediate designs are feasible but only that the optimum (if found) is feasible. The advantage of SQP is its generally increased speed over GDP. For further information on these algorithms, and optimization in general, consult the excellent text Introduction to Optimum Design by J.S. Arora (McGraw-Hill, 1989), Chapter 6.

We will have occasion in the following lessons to experiment more with the settings for an optimization. For now, let's examine what happened with our bar.

Open the **Study Status** window and browse through the run report. Notice that the maximum stress constraint is violated for the initial design values. It takes two iterations to find a design that satisfies the stress limit by increasing the section width (and the part mass). The values used for the design variables are given. Then it proceeds to minimize the mass while satisfying the stress limit by increasing the bend radius up to the maximum allowed value. Note the final optimized values for the variables are (width = 26.8 mm, bend_radius = 60 mm). The mass has actually increased from the initial value, but remember our starting point violated the maximum allowed stress condition. At the bottom of the report, note that the total number of analyses was 19 - another reason to have an efficient model. When the run is finished, close out the result window.

Optimization Results

Use *Results* as usual to create some result windows. Remember to select the design study **bar_1opt** as the source of the data to be plotted. Set up a window to show the Von Mises stress fringe plot. This will show the stresses in the final optimized design. See Figure 8. Note that the maximum reported stress is 35 MPa as required[4].

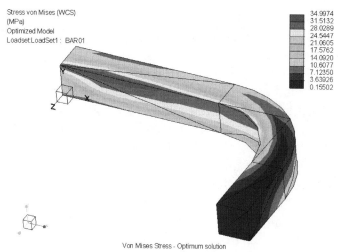

Figure 8 Von Mises stress on optimized geometry

Also, create two windows to plot the measures **max_stress_vm** and **total_mass**. See Figure 9. These will automatically be plotted as graphs, with the horizontal axis being the optimization iteration. Note the condensed vertical scales. The initial design (pass 0) violates the limit on the Von Mises stress. Two passes were required to reduce the stress to an acceptable level. In order to do that, as seen on the figure at the right, the mass of the part had to increase. As of pass 2, the design was feasible. On the next passes, the optimization was able to reduce the mass somewhat, while maintaining the stress limit.

---

[4] Where does the maximum stress occur? What would happen if we had used the user-defined measure **maxVM_fillet** for the constraint instead? This would be a better idea.

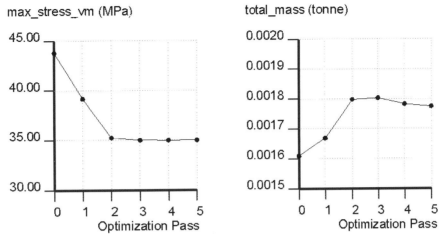

Figure 9 Optimization history: Von Mises stress (left), total mass (right)

The values of the design variables during the optimization iterations are shown in Figure 10 (this graph was not generated by Creo). This data is stored in the file *bar_1opt.dpi* in the results directory.

We are finished looking at result windows for now. Return to the **Analyses and Design Studies** main window.

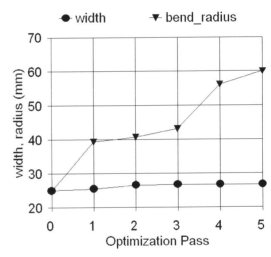

Figure 10 Variation of design variables during optimization

Viewing the Optimization History

Notice that the display of the part in the Creo window is showing the original geometry. We want to transfer the values of the design parameters for the optimized geometry back to the part itself. At the same time, we can view a history of the shape changes during the optimization process given by the variable changes shown in Figure 10.

In the **Analyses and Design Studies** window make sure the study **bar_1opt** is selected[5], then in the pull-down menu select

Info > Optimize History

---

[5] You may have to move the **Analyses and Design Studies** window out of the way so you can see the model.

From here, follow the message prompts. The part is first shown in the original geometry, with the initial values of the design variables (both 25). You can now step through the optimization passes and the part will regenerate with the new design variable values. Use middle click to advance to the next shape. When you reach the final shape, Creo asks if you want to leave the model at the optimum shape. Select *Yes* or middle click. The model now has the optimized design parameters. Leave Creo Simulate by selecting *Close* in the Home ribbon and check the dimensions with the *Edit* command. These are shown in Figure 11. We need the original part **bar01** a bit later, so use *File > Save As > Save A Copy* and save the optimized part with the new name **bar02**.

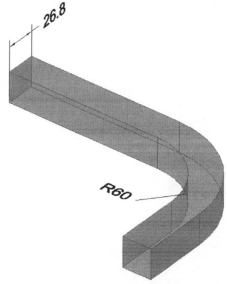

Figure 11 Optimized design

This optimization was pretty simple because the two design parameters didn't affect each other very much. The maximum stress was basically determined by the section width, and the total mass by the bend radius. Our optimized design, in fact, was determined by the maximum allowed bend radius (in optimization terms, the constraint was *active*). We could possibly have guessed that this would be the case, and saved ourselves the trouble of doing the optimization run. However, in other problems, the interaction of the design variables may be much more difficult to predict. For example, did you expect the location of the point of maximum Von Mises stress to move?

As a final note, in a more complicated design problem, there may be several local optimum solutions surrounding a global optimum. The optimization process used in Creo (as in many optimization routines) does not guarantee that the global optimum will be found. If this is really what you are after, it is necessary to have a good idea where the global optimum is, and restrict your search (by setting the starting points and ranges of the design variables) to a region close by.

Considerations for Applying Loads and Constraints

Up to now, we have been careful to apply loads and constraints on surfaces of the part. Let's do an experiment to see what happens if we deviate from this.

Erase the current model (the optimized **bar01**), and load the original (non-optimum) part **bar01**. (If you don't have the original, just use *Edit* and change the **width** and **bend_radius** values back to 25 and *Regenerate*.) Transfer into Creo Simulate and open the model tree. Expand the **Loads/Constraints** entry, and use the mini toolbar command to *Suppress* the named constraint **fixed_face** (this also suppresses the related reaction measures there) and named load **endload**.

In the **Constraints** group, select:

Displacement

Set up a displacement constraint *[fixed_edges]* by fixing all degrees of freedom of only the top and bottom edges on the left end. In the **References** pull-down list, select *Edges/Curves*, then pick on the upper and lower edges of the face we constrained before (use CTRL-left to pick the second edge), then middle click. This leaves everything fully constrained. You should see the constraint symbols on the edges. If you forgot to specify a name, you can do that by selecting the constraint in the model tree or graphics window and use the *Rename* function in the RMB pop-up menu.

We are going to apply point loads to the top two corners on the other end of the bar (half the load on each). In the **Loads** group, select:

Force/Moment

Name the new load *[point_loads]*. In the **References** list, select *Points*. Let's create a couple of datum points here[6]. We can do that "on-the-fly" by going to the **Refine Model** ribbon and selecting *Point* in the **Datum** group. Create two datum points on the end of the bar (see Figure 12), then middle click. Points PNT0 and PNT1 will be displayed (unless that has been toggled off). Complete the Force/Moment dialog window by entering an X-component of **500** and a Y-component of **-250**. Note that the load is automatically split between the two points, a consequence of the *Total Load* option that was the default setting under the *Advanced* button. Accept the dialog with a middle click. The model should now look like Figure 12[7].

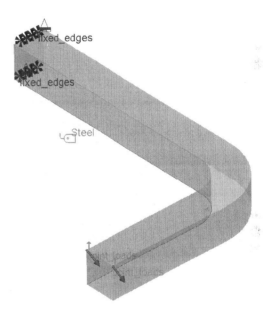

Figure 12 Edge constraints and point loads

Before proceeding, to verify that the same total load has been applied, in the **Loads** group pull-down menu select

Review Total Load

---

[6] This is not strictly necessary here, since we could pick the two vertices at the corners, but reminds us of the ability to create datum features on-the-fly. We could, for instance, create a datum point on the center of the square end face where there is no vertex.

[7] To modify the display, select *Simulation Display* in the Graphics toolbar.

Click the button under **Loads**, then on one of the load arrows to identify the load, middle click, then select ***Compute Load Resultant***. The resultant magnitudes of FX and FY are exactly what we used before. The moments are measured at the specified evaluation point (default: origin) using directions defined by the indicated coordinate system (default: WCS). Close this window.

Create a new static analysis called ***[bar_1d]*** ("d" is for diabolical!). Enter a description and make sure the constraint and load sets are highlighted. Set up a **Quick Check** convergence.

Now ***Run*** the analysis **bar_1d** and view the interactive diagnostics window shown in Figure 13. You will see some warning messages about the point loads and edge constraints. If you click on these messages, the graphics window will zoom the display in on the offending items. Note that these are warnings only and not errors. That is, these problems do not stop the model from executing. Whether or not the results are reasonable is another matter, as we shall see soon. Note that one of the warning messages indicates that some of the reported results must be considered very carefully, and will be indicated by an asterisk in the report summary.

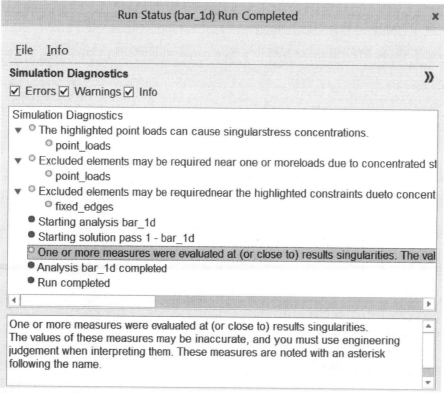

Figure 13 Simulation Diagnostics window showing errors arising from point loads and constraints

Open the ***Summary*** window and see what messages are in there. There is a message near the beginning about "excluded elements", which are discussed below. In the Measures area at the bottom of the report several measures are marked with asterisks. What do these all have in common? They are all *stress* values. Recall that while displacements are the primary solution variables, stress is a derived result that will amplify errors in the solution. A warning message indicates that these "were evaluated at or close to" singularity locations and "you must use engineering judgement when interpreting them." Other than this, there is not much indication of the serious problem coming soon (for instance, check out the maximum Von Mises stress which is not much different from the previous result but with no hint of convergence properties!). Close the **Study Status** window.

Now *Edit* the analysis **bar_1d** to run a **Multi-Pass Adaptive** analysis with a maximum order 9 and convergence criterion of 5% on **Local Displacement, Local Strain and Global RMS Stress** as before. *Run* this analysis and look in the *Study Status* window. You will find that the analysis will not converge after all 9 passes. The maximum displacement is a bit larger than we had before, 0.64mm. However, the reported maximum Von Mises stress (>300 MPa) is many times larger than before! Every reported stress value is marked with an asterisk. Clearly, there is something peculiar about these results.

To investigate this, create result display windows to show the Von Mises stress fringe plot, and the convergence behavior of the Von Mises stress, maximum displacement, and total strain energy. You may have a result template for this.

Show the Von Mises stress fringe plot. Observe first that the legend is linearly distributed between zero and the maximum stress (in fairly large increments). This reveals the "hot spots" at the locations of the applied loads and constraints, but doesn't tell us much about the rest of the model, which is all the same fringe color (figure 14). The values of our previous model would have been all included in the lowest couple of fringes. You can select *Format > Edit* to change the values attached to the color fringes. If you have an 8-level legend, set the minimum value to 5 and the maximum value to 40 (see Figure 15). After doing this, you will see that, for the regions far from the applied loads and constraints, the fringes look pretty much the same as the previous model.

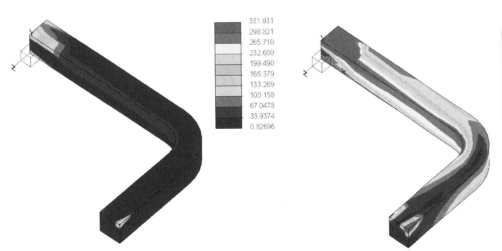

Figure 14 Von Mises stress in the diabolical model

Figure 15 Von Mises stress in the diabolical model (modified legend)

The effect of the applied point loads is quite evident. This is what we were warned about before. Notice the effect of the applied constraints (on the edges at the other end of the bar). These are also causing some stress concentration in the model very close to the constraint. The reported stresses in these areas are very dubious. What is the location of the maximum stress (use **Model Max**)? Is this the same as before? What is the value on the fillet (use **Dynamic Query**) and how does it compare to the previous model?[8]

What is causing this behavior? It is easy to visualize that a force applied on a point (of zero area) will cause an infinite local stress and therefore extremely high gradients of stress as you move away from the point. What Creo is trying to do during the multi-pass adaptive analysis is to capture this infinite value and very high gradients. The only way it can do that is to continually increase the element order. If we were to let it go, it would do this indefinitely, since it is impossible to capture this "infinite" value. That is why there is a maximum allowed edge order. Such a point in the solution field is called a *singularity*. The same effect occurs at the location of the constraints in this case. ***The maximum reported stress in this model is associated with the singularity arising from how the loads and constraints are implemented in the FEA model, not with any real physical behavior.*** Clearly, this modeling behavior is something we should be on the lookout for and is the reason for the warnings we saw earlier.

Edit the stress fringe result window to show **Quantity (P-level)**. You should see that elements with edges or vertices associated with the point loads and edge constraints are all at the 9th order. The presence of 9th order elements should therefore be cause for alarm.

If you are not particularly interested in what is happening very close to a singularity, these results might be all right. For example, if you are interested only in the stresses in the vicinity of the bend in the bar, then you wouldn't worry.

However, the presence of these numerically induced stress concentrations has seriously disrupted the convergence monitoring within Creo. Figure 16 (right) shows the convergence behavior of the maximum displacement. The strain energy graph has the same general shape. This does not show movement towards a constant value. Figure 16 (left), which shows the maximum Von Mises stress convergence behavior, is a clear indication that something is wrong with the model. This is the type of graph you can expect to see if there are singularities in the model - the model maximum Von Mises stress will not settle down to a constant value, no matter how many p-loop passes you take, and appears to be increasing indefinitely. ***This effect is not changed by increasing the mesh density, although the divergent behavior may be delayed to later p-loop passes.*** For this reason, by default the convergence monitoring routines in Creo do not use the model maximum stress but rather a global averaged stress value. However, if you have a simple model with not many elements in it, as we do here, the effect of a singularity will be large enough to obscure or mask the convergence of the rest of the model. This causes the MPA to go to continually higher orders, *ie* blow up!!

---

[8] This is an illustration of a rather important principle in strength of materials regarding what happens far removed from the point of application of loads or constraints. Can you remember whose name is associated with this principle?

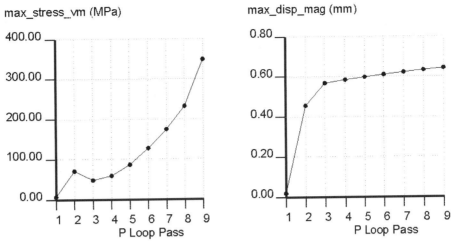

Figure 16 Convergence behavior of displacement and Von Mises stress in diabolical model

A simple solution to this (although not the best) is to not request excessively tight convergence tolerances. For example, if we had requested a 10% convergence on default measures, the MPA passes would have stopped on pass 4 (Try it!). In that case, would you say the solution had actually converged or just stopped iterating?

The lesson to be learned here is that, to avoid convergence problems for solid models, if you can manage it, **you should not apply loads or constraints to points or edges - only surfaces**. We will encounter the same type of restriction in other models we will see later. This is a direct result of the p-code method and the internal convergence monitoring process used by Creo. If you really want to constrain or put a load on a very small area, you can do this by defining a small region on the surface of the model using a datum curve, and applying the load/constraint on the region. This reduces but does not totally eliminate the problem.

If point (or edge) loads or constraints are unavoidable, there are a couple of ways we can set up the model. In some later lessons, we will investigate alternate methods for monitoring convergence using custom measures that are not affected by a singularity. There is also a very easy to use tool which is available to deal with singularities: excluded elements. These are discussed in the next section.

Using Excluded Elements

A singularity involves a very localized effect within an element that is adjacent to the singular location (point or edge load or constraint). Such an element must resort to very high order polynomials to capture the high gradients and infinite stress that occur near and at the singularity. In a model with few elements and especially where many of those elements are adjacent to the singular locations (as we have with our bar), the effect of these elements on the convergence analysis will be strong. That is, although the solution in much of the model is satisfactory, the adaptive convergence analysis is mislead by the corrupted data into continuing to higher orders - a process that cannot actually result in true convergence.

To get around this problem we must do two things. First, in AutoGEM, we identify the singular points (this is automatic) and specify how we would like the mesh to be modified (primarily the element size) in the vicinity of these points. This control will cause AutoGEM to create a number of additional elements in direct contact with the singularity. Then, when setting up the design study, we will cause those elements (and the results obtained on them) to be ignored during the convergence analysis. Collectively, performing these steps is called *excluding* the elements. Let's see how this works.

The model should be in the configuration shown in Figure 14 - point loads and edge constraints. To create the AutoGEM control, go to the **Refine Model** ribbon and in the **Control** pull-down list (Figure 17) in the **AutoGEM** group, select

Isolate for Exclusion

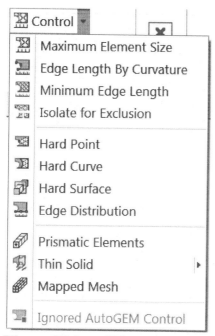

Figure 17 AutoGEM Controls

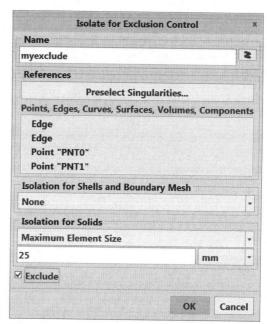

Figure 18 Setting the AutoGEM control for excluded elements

In the window that opens, Figure 18, enter a name like **myexclude**. Now select the button *Preselect Singularities*. In the new window that opens, make sure the boxes beside point loads and edge constraints are checked, then *Select*. These now show in green on the model in the graphics window. All entities are listed in the collector in the Exclusion Control window. In the **Isolation for Solids** pull-down list, select *Maximum Element Size*. Enter a value of **25**. This tells AutoGEM how large you want the elements to be that contact the singularity locations. We will come back later and explore the effect of this setting. For now, note that we have chosen a size basically the same as the cross section of the bar. The completed dialog window is shown in Figure 18. Select *OK*.

We now see new icons on the screen (see Figure 19) that indicate the control. The resulting mesh (use *AutoGEM > Create*, then shrink the elements) is shown in Figure 20. In the AutoGEM dialog window, select

Info > Isolating Elements

and the elements to be excluded will highlight in red.

In the mesh (which now contains over 50 elements instead of the 21 or so previously), you can see a number of additional elements have been created adjacent to each point and edge where there is a control. The longest edge on an element touching the singularity is about 25

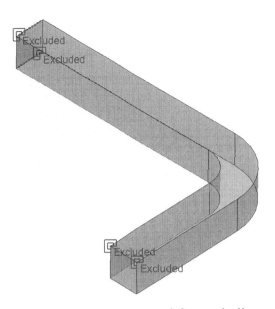

Figure 19 AutoGEM control for excluding elements

mm, as specified. This means that the closest adjacent elements are about this far away from the singularity. The idea is that, when we later exclude these elements contacting the singularity, the remaining elements will be far enough away that they will not be significantly corrupted. If we make this distance too short, some of the effect of the singularity may still be felt in the adjacent elements; if we make it too long, we will be excluding perhaps too much of the model from the convergence analysis. So, once again, we are faced with a situation that may require compromise in setting up the model.

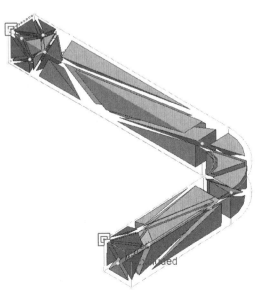

Figure 20 Mesh created with controls of Figure 18

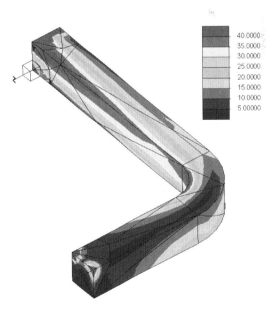

Figure 21 Von Mises fringe plot of model with excluded elements and MPA

Now we can return to the **bar_1d** design study. Keep the MPA convergence on default measures with 5% convergence with a maximum order of 9. Select the tab for *Excluded Elements*. Turn on the check boxes beside **Exclude Elements** and **Ignore (Stresses)**. To try to prevent the suspect elements from "blowing up", set the *Limit* on the **Polynomial Order** to **6** (note that this only affects the excluded elements). Accept the run settings and start the run.

The MPA will go to pass 8. In the **Summary** window, the **max_disp_mag** is 0.69mm. The reported maximum Von Mises stress is about 57 MPa - somewhat higher than before. We'll have to investigate this. Create the usual result windows.

The Von Mises stress fringe plot is shown in Figure 21. Adjust the fringe legend levels (levels 8, min 5, max 40). For most of the model, these fringes agree with previous results. The legend shows a model maximum of several hundred MPA, which occurs at one of the point loads. This is much higher than what is reported in the **Study Status**, which indicates that the latter is not considering the excluded elements (also indicated by absence of asterisks). If you do a *Dynamic Query*, you can explore the rest of the model. The maximum stress at the inside of the bend is what we obtained before.

The convergence graphs for this run are shown in Figure 22. Note that the displacement has converged very well (although at a higher value than the original model). The maximum Von Mises stress has also converged quite well but not to the same value as before. Where does this maximum stress occur? Show the Von Mises fringe, then select

Home { Report } : Measures

and select the measure **max_stress_vm**. A small circle will appear on the model at the corresponding location. This is not at the singularity (or on the bend) but on the closest non-excluded element. This is a bit of a surprise, and points out again that you have to be careful in interpreting the Creo output. The stress data shown in Figure 22 (or in the Summary) does not relate to what is happening at the bend.

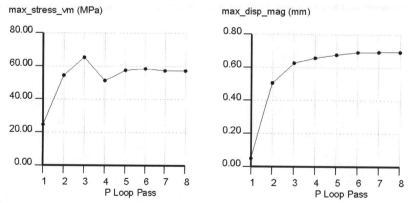

Figure 22 Convergence of model with excluded elements (maximum size = 25 mm, maximum polynomial order = 8)

Leave the results windows and pick the AutoGEM control icon in the graphics window or model tree. In the mini toolbar, select **Edit Definition**. To increase the distance of the adjacent elements from the singularity, increase the maximum element size from 25 to **50** mm. In the Design Study definition window, go to the **Excluded Elements** tab and change the **Limit** on the polynomial order of excluded elements to **4**. These settings produce the convergence graphs shown in Figure 23. This model converges much quicker. The maximum Von Mises stress is indicated as 49 MPa - much closer to our previous results, but you should check where this actually is happening.

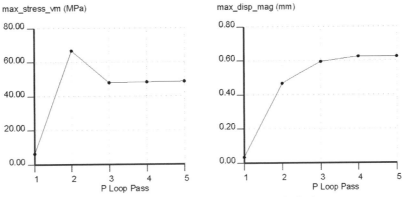

Figure 23 Convergence of model with excluded elements (maximum size = 50 mm, maximum polynomial order = 4)

Superposition and Multiple Load Sets

In the problems treated so far, we have provided a single load set to describe the total load on the bar. Suppose that you wanted to analyze the performance of the design under many different loading scenarios but with the same constraints. Do you have to analyze each problem separately? This would obviously involve a possibly large number of models and extensive computer time. Fortunately, the answer to this question is "No!" ... read on!

In the solution of our bar problem, the governing equations for the static stress solution are linear, the material properties are assumed to be linear, and the geometry does not undergo a large deformation. Therefore, our entire problem is linear. You probably know from calculus that for linear problems, you can combine different solutions (this is *superposition*) and the combination will also satisfy the problem (governing equations and boundary conditions). This is a very powerful concept in FEA. In this section we will see how to set up a solution for a multiple of applied loads, and how to superpose the various solutions to analyze an infinite number of loading possibilities. Surprisingly, this does not require a huge increase in computer time.

To get started, bring in the original part **bar01** that we used previously. If you need to, delete or suppress any point loads, edge constraints, AutoGEM controls, and datum points (if necessary) - do this from the model tree. We previously ran this model with a load FX=500 and FY=-250 applied to the face on the end of the bar. Suppose we wanted to examine the behavior of the bar for many different values and combinations of FX and FY. This is where we can use multiple load sets and superposition.

The left face of the bar should be constrained as before (all degrees of freedom **FIXED**). If you still have the previous **fixed_face** constraint in the model, just resume it. A surface constraint will avoid the singular behavior we saw in the previous section.

Creating Multiple Load Sets

We are going to separate the load on the end of the bar into components in the X and Y directions, and apply them separately. Start with the *Force/Moment* button. Beside the **Member of Set** pull-down list, select the *New* button and enter a name *[Xload_set]*, then *OK*. Enter a load name *[Xload]*. Click on the end face of the bar. Middle click. Enter an X component of **100** and *OK*.

Repeat this procedure to create a new load set called *[Yload_set]* containing a load *[Yload]*. Apply this load to the same surface and enter a value for the Y component of **100**.

It doesn't really matter what values we enter for the load components here, since when we combine the loads later we will be multiplying each load by a scaling factor. You could, for example, enter unit loads here. If you do this, the scaling factors become the load magnitudes. You can even reverse the direction of the load by using a negative scaling factor.

The model now should look like Figure 24. Expand the model tree to see the two load sets.

Setting the Analysis for Multiple Load Sets

We have to set up an analysis definition to tell Creo to use both load sets. In the **Analyses and Design Studies** window select

> *File > New Static*

Enter the name of the analysis as *[bar_1m]* (m is for multiple). Click on both load sets (**Xload_set** and **Yload_set**) for the analysis (you may also have to deselect LoadSet1, or better yet, delete it). The button **Sum Load Sets** will cause the two load sets to be logically merged in the run - not what we want to do here. Set a **Multi-Pass Adaptive** analysis with a maximum polynomial order of 6, with 10% convergence. The definition appears as shown in Figure 25. Accept the definition with *OK*.

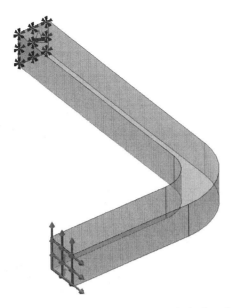

Figure 24 Model with multiple loads

Figure 25 Creating an analysis with two load sets

In the **Analyses and Design Studies** window, select the analysis **bar_1m**. Check the run settings. Since we turned off the AutoGEM controls we can't use the existing model mesh. Check the option to **Override Defaults** and the option **Create elements during run**, then you can *Start Run*. While the study proceeds, open the *Study Status* window. Near the bottom of the file, you will see separate results for the two load sets. For load set **Xload_set**, note the maximum displacement is 0.083mm. There is some displacement in the Y direction (-1.4E-04mm), which is not expected here due to symmetry of the geometry and the load **Xload**.[9] It is possible that our convergence criterion is not tight enough to bring this to zero. For the **Xload_set**, the maximum Von Mises stress is 6.7 MPa. For load set **Yload_set**, the maximum displacement magnitude is 0.14mm and the maximum Von Mises stress is 10 MPa.

Combining Results for Multiple Load Sets

First, we'll create two result windows that will show fringe plots of the stress due to each load set separately.

---

[9] You might like to create a result window to show the Y displacement, to find out where this value occurs. Is it possibly related to the mesh (which is not symmetric about the mid-horizontal plane)?

Create the first window with a name *[vmx]*. Enter the title *[Von Mises - Xload]*, or something similar. Make sure the design study is **bar_1m**. The **Result Window Definition** will show the two load sets. See Figure 26. Select only the **Xload_set** and leave the Scaling set to 1. Keep a fringe plot. In the **Display Options** tab, turn on the display of element edges. Accept the window definition with *OK and Show*.

Copy this result window to another window, **[vmy]**. This time, select the **Yload_set** only. Once again set up a fringe plot and accept the dialog. Show the two result windows **vmx** and **vmy**. These are shown in Figure 27 below.

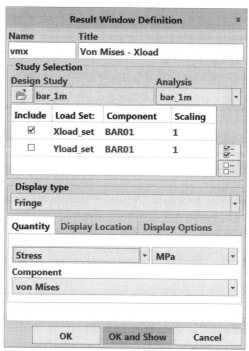

Figure 26 Selecting results for a single load set

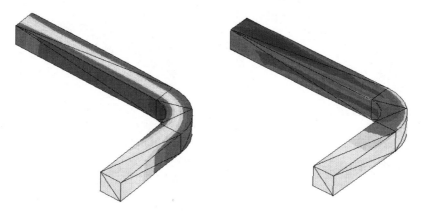

Figure 27 Von Mises stress fringes for Xload (left) and Yload (right). The fringe spectrum has been modified for these images (use *Thermal*, inverted)

Now create a new window called *[vmcombined]*. Keep the same design study. Include both load sets. To create the same load conditions that we had previously, the scaling factors should be **5** for **Xload** and **-2.5** for **Yload**. The result window (Figure 28) shows the Von Mises stress fringes for the combined loads. This can be compared with Figures 28 and 29 in Lesson 3. Check the maximum Von Mises stress and compare it to what we had before. Minor variations are probably due to the change in the convergence criterion and the number of solution passes for the present results.

Copy the definition **vmcombined** to a new window *[vmcomb2]*. Enter a new title ("Von Mises combined #2"). Make the **Scaling** for the Xload **2** and for the Yload **-5**. Accept the dialog and show the window.

Copy the definition **vmcomb2** to a new window *[def2]*. Set this definition up to plot an animation of the deformation.

The windows **vm2** and **def2** are shown in Figures 29 and 30 below. The results are consistent with our expectations.

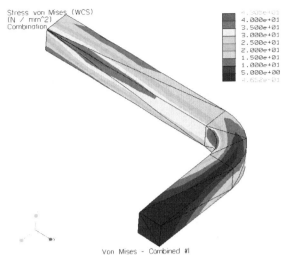

Figure 28 Von Mises stress for combined load (5*Xload, -2.5*Yload)

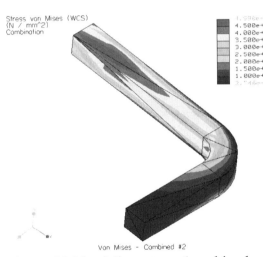

Figure 29 Von Mises stress (combined loads #2)

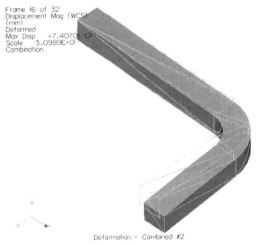

Figure 30 Deformation under combined loads #2

You see how easy it is to set up the model to handle a multitude of possible loading conditions just by changing the scaling factors in the result window. Each of our load sets here had only a single load defined in it. It is possible to put several different loads into a single load set (a force and a pressure, for example), but these cannot be separated later. The scaling factor in the results display definition would apply to all loads in the set.

This method of treating multiple loads (with the same constraints) is very efficient, since the problem must only be solved once. Adding load sets does not appreciably increase the solver time. After the solution, various load combinations are put together by post-processing the data, which is a very simple computation. Even if you think loads are related, it does not hurt to put them in separate load sets. This does not cost you much, and gives you a lot of flexibility in result interpretation. The use of multiple load sets and superposition is probably one of the most underutilized capabilities in FEA.

Summary

This lesson has covered quite a lot of ground. We looked at the main steps involved in setting up and running standard and sensitivity studies. The main item of interest here is the selection of design variables, which is very easy to do. Design variables are also the key ingredient in performing optimization. For both sensitivity studies and optimization, it is necessary to have a pretty good idea of how the design variable is used to control the geometry, and the effect it will have on the model. This is particularly true when more than one design variable may interact, possibly causing an illegal geometry.

Optimization can become a very complex problem. Your safest approach is to have some idea of the solution, and keep your search range small. It also helps to keep this process in mind when you are creating the geometry in Creo Parametric. Try to keep your parts as simple as possible, and you will avoid unpleasant surprises!

We also saw the effects of applying loads and constraints to points and edges of a solid model. The lesson here is - don't do it! Point and edge loads and constraints on solid models will lead to singularities in the solution that will interfere with the normal convergence process (and give erroneous results in some regions of the model). We saw how to set up an AutoGEM control for automatic detection of problem locations, and specify mesh adjustments in those areas. We also saw how the design study can be set up so that elements at these locations are excluded from the convergence analysis.

As we will see later, singularities can also occur within the model itself (not directly associated with loads or constraints). For example, sharp concave or "re-entrant" corners are often problematic. These will also interfere with the monitoring of the convergence of the model. Although stress values reported at singularities are meaningless, the values in the rest of the model are probably OK. In subsequent chapters we will examine some more methods to practically deal with the presence of singularities in the model.

Finally, we looked at the use of multiple load sets. These can save a lot of time if a model must be analyzed under a wide range of applied loads.

This basically concludes our look at solid models and all the main tools and functions in Creo Simulate. In the next few lessons we are going to look at how idealizations can be used to solve some model geometries much more efficiently. We will also experiment with additional AutoGEM controls for fine-tuning the mesh to create efficient and accurate models.

Questions for Review

1. Give an outline of the necessary steps in performing a sensitivity study.
2. How do you specify a design variable and its range?
3. What will generally happen to the convergence behavior when you produce a finer mesh in Creo? Why?
4. When setting up the design variables, how can you determine the geometric effect of changes in a single variable?
5. When running a sensitivity study, what does the *"repeat p-loop convergence"* option do? Under what circumstances would or wouldn't you use this?
6. What is the *shape history*, and how do you obtain it?
7. What should you always do after creating one or more design variables?
8. What search algorithm does Creo use in optimization?
9. How can you plot the variation in a measure that occurred during the optimization?
10. How do you save the final optimized design?
11. What information is required to set up an optimization design study?
12. How do you carry the optimized geometry determined in Creo Simulate back into Creo Parametric?
13. What is the potential problem with the use of point and edge loads and constraints?
14. A model can have several load sets. Each load set can have a number of load definitions. How many constraint sets and individual constraints are allowed?
15. What model condition is required in order to use multiple load sets?
16. How many loads can be set up within a single load set?
17. What is St. Venant's Principle, and why is it relevant to FEA?
18. How can you automatically detect potential singularities in the model?
19. How can you make sure elements touching singularities do not overly affect the solution in the rest of the model? What is this process called?
20. In the Measures table of the Result Summary window, what does it mean when there is an asterisk beside a reported value?
21. How can you find the location on the model for the reported measures in the Study Status window?

Exercises

1. Using the bracket model from Exercise #5 in the previous lesson, perform a sensitivity study to examine the Von Mises stress as the radius of the fillet varies from 0.25" up to 1.0".

2. Optimize the bracket from Exercise #5 in the previous lesson to yield a minimum total mass. The design variables are the thickness of the base plate (minimum 0.25", maximum 2.0") and upright (minimum 0.5", maximum 1.5"), and the radius of the fillet (minimum 0.25", maximum 1.0"). The Von Mises stress should be limited to 1500 psi.

3. Using the connecting rod (Exercise #4 in previous lesson) (or the hook, Exercise #1), modify the applied loads to use multiple load sets.

4. A solution involving superposition, judicious defeaturing, selection of constraints.

Download the file ***rodend.prt*** from the SDC web site. Assuming that the material is an annealed steel with $S_y = 450$ MPa, determine the maximum permissible load "F_i" for the three load cases listed in the table. Assume that the threaded end is fully inserted into a fixed and rigid base as far as the shoulder. Use a factor of safety of 2. Where is the point of failure in each case? As always, think about the quality of your solutions (e.g. convergence) and how you can set up the model to produce reliable results. Consider carefully how you might defeature the solid geometry, and how to constrain the model.

Load Case, i	Axial Load	Vertical Load	Side Load
1	F_1	$0.2\,F_1$	$0.2\,F_1$
2	$0.2\,F_2$	F_2	$0.1\,F_2$
3	0	$0.3\,F_3$	F_3

HINT: Apply each individual load as a unit load. Do the load superposition and magnitude scaling when you create your result windows. You should be able to get the desired results all from a single run of the solver (after you have properly set it up).

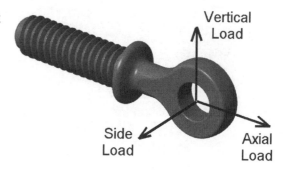

5. Setting up and running an optimization study

The figure below shows a conceptual design for a wall bracket that is to support a bearing load F = 25000N directed as shown (45 degrees below horizontal). A manufacturer plans on producing thousands of these brackets, and so material requirements are important. The layout of mounting holes on the back plate is fixed, since these must line up with existing bolts which cannot be moved. However, the dimensions shown in the figure are variable. Find an optimum design of the bracket (values for the indicated dimensions) that will minimize the amount of material used. A factor of safety of 2.5 is required and the steel used has a yield strength of 325 MPa. Limits on the values of the dimensions shown are:

	Minimum	Maximum
bracket thickness	10 mm	60 mm
back plate thickness	10 mm	60 mm
fillet radius	5 mm	15 mm

The bracket part **bracket.prt** can be downloaded from the SDC web site.

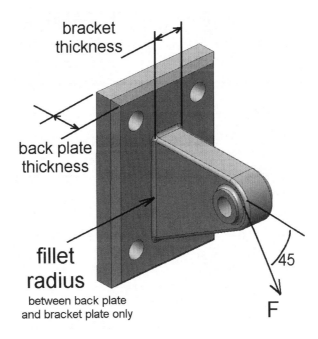

NOTE: This optimization run could take a while, depending on how you specify the static stress analysis design study and the mesh you use. A little experimentation here would be useful to make sure you have an efficient solution. (Hint: where is the critically stressed element, and can you defeature the model to reduce its complexity without affecting the desired results?)

6.   Perform a static stress analysis of the L-shaped bar shown below for the load **F =
     1000N** in the direction indicated. Dimensions are in **mm**. This is essentially the
     same geometry as used in the last two lessons (in a different orientation), with the
     main difference being the circular cross section. This will allow you to compute *by
     hand* all the stresses at points **A, B, C**, and **D**, using the formulas given in Exercise
     3 of Lesson 1. Make sure to use multiple load sets so that you can easily find the
     separate contributions due to bending, torsion, axial load, etc., to compare with the
     hand calculations. Comment on the similarity or otherwise of the results of the hand
     computation and the FEA.

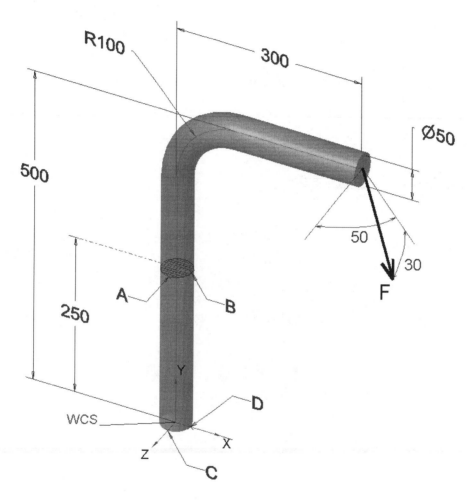

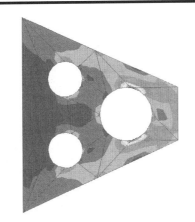

Chapter 5 :

Plane Stress and Plane Strain Models

Overview of this Lesson

As demonstrated in the previous two chapters, the default model type is a 3D solid, for which the finite element model is composed of tetrahedral elements. For many FEA problems, however, treating them as solids is "overkill." Useful results can often be obtained much more efficiently by treating a simplified version of the problem using *idealizations*. For example, the essence of the problem might be expressed using only a two dimensional planar geometry. This is the subject of this lesson. Other idealizations (shells, axisymmetric solids and shells, beams) are treated in the next few lessons.

Although Creo is by nature a 3D modeler, you can easily treat 2D FEA problems. These include models for plane stress and plane strain based on the geometry contained in the 3D model. For plane stress and plane strain, as the name implies, the geometry of interest is contained in a planar surface that is "extracted" from the 3D model. The solution is obtained using planar elements (quadrilaterals and triangles) instead of tetrahedral solid elements.

For all problems based on 2D geometry, some restrictions and additional points need to be considered:

1. All surfaces used to define the geometry for the FEA model must be co-planar. No surprise there, but what this means is you cannot project another surface onto the FEA model plane.
2. The geometry must have an associated Cartesian coordinate system. This can be created in Creo Simulate and does not have to align or be parallel with the default or WCS axes - it can be any orientation in 3D space.
3. All entities for the model must be in the XY plane of the coordinate system. This includes all model geometry, constraints, and loads.
4. For axisymmetric models (these are also 2D models), the model must all be on the positive X side of the coordinate system, X > 0.

In this lesson, we will examine plane stress and plane strain models. Axisymmetric problems are also based on planar models, but we will defer these to the next lesson.

We'll also explore the use of symmetry to further reduce the size of the model (and hence its computational cost). The use of simulation features (coordinate systems, datum curves, regions) is also introduced as well as pressure and temperature loads. These features are also useful for 3D models. And as usual, there are Questions for Review and some suggested Exercises at the end.

Plane Stress Models

A plane stress problem arises when a model is very thin in one dimension compared to the other two. A typical example is a thin flat plate. Because the part/model must stay in the same (XY) plane, loads must also be applied only in the plane of the plate[1]. In this case, the normal stress σ_{ZZ} (perpendicular to the plate) is assumed to be zero throughout its thickness[2]. Normal stresses σ_{XX} and σ_{YY} are in the plane of the model and are constant through the model thickness. The equations that govern plane stress problems are considerably simplified, and plane stress models are among the fastest to compute.

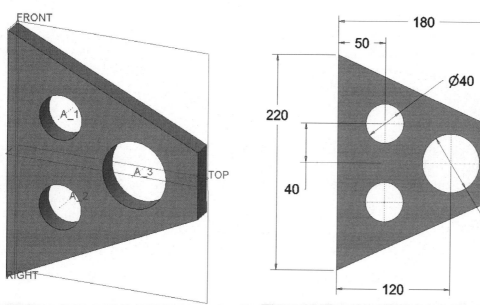

Figure 1 Flat plate solid model

Figure 2 Flat plate dimensions in mm (20 thick)

The example problem we will study concerns the stress analysis for the thin flat plate shown in Figure 1. Create this model (call it **psplate**) in Creo as follows and according to the dimensions in Figure 2. You might like to put this in a new directory ("Chap05"). Use a solid part template and make sure your units are set to mm-N-s. The model consists of a solid protrusion sketched on **FRONT** and some straight holes. Create the lower hole as a mirror copy of the upper hole. To illustrate a point with 2D plane stress

---

[1] If there are out-of-plane loads, the model must be treated using shells, which are discussed in Lesson #7.

[2] In fact, all stress components in the Z direction are zero.

modeling here, create the solid plate with a thickness of 20mm. You can delete the default coordinate system.

When the Creo part is complete, launch Creo Simulate using

Applications > Simulate

The default is to create 3D solid model. We will come back in a minute to change this.

Note that the green WCS is created automatically at the default location. As mentioned previously, a 2D model must have an associated Cartesian coordinate system whose XY plane contains the surface geometry for the analysis. We are going to use the front surface of the plate to define our 2D geometry, so we'll need to create a new coordinate system as a *Simulation Feature* there.

Creating a Coordinate System

To create the new coordinate system, go to the **Refine Model** ribbon and in the **Datum** group select

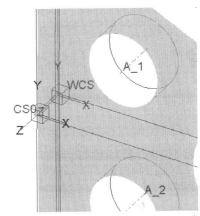

Coordinate System
Type(Cartesian)

Holding down the CTRL key, click on the front face of the plate, the **TOP** datum, and the **RIGHT** datum. As they are selected they will be listed in the **References** collector in the **Coordinate System** window. A new system will be created at the intersection of these three planes. This is probably not in the desired orientation. If not, pick the

Figure 3 Coordinate system CS0

Orientation tab. Select the appropriate surfaces and directions (X, Y, or Z) and use the *Flip* button as necessary to obtain the coordinate system **CS0** shown in Figure 3 (there are several combinations to choose from). For example, use the surface reference to define the Z direction, and the TOP datum to define Y (both directions are normal to the chosen reference). Open up the model tree to see an entry for the feature there (under **Simulation Features**). When you go back to view the part in Creo Parametric, this feature will not appear in the model - it is known only to Creo Simulate.

Setting the Model Type

Now we can start setting up the plane stress model. In the **Home** ribbon, select

Model Setup > Advanced

This brings up the **Model Setup** window that we have seen before (Figure 4). Check the button beside **2D Plane Stress**. With the **Coordinate System** collector active, select *CS0*. The **Geometry** collector then becomes active (if not, click inside the collector), so select the front surface of the plate. It will highlight in green. Middle click (or *OK*).

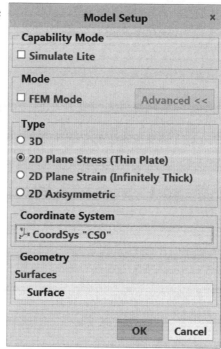

As you leave the **Model Setup** window, a warning message will let you know that if the model type is changed (remember the default was a 3D solid), all previously defined FEA modeling entities will be deleted (loads, constraints, materials, and so on). If that happens, you would have to create them again. Thus, be very careful about selecting the model type, since if you pick the wrong one and have to come back and change it later, much of your work will be lost. Click *Confirm*.

Figure 4 Changing the model type

Turn off all the datum axes, planes and so on to remove some of the screen clutter. The 2D surface is now highlighted in purple (this is easier to see in wireframe).

Applying Loads and Constraints

You can now apply loads and constraints in the same way as we did previously. Since we are dealing with a planar model, loads and constraints should be applied to edges (rather than points, which may lead to convergence problems). We will fix the left edge of the geometry and apply a uniformly distributed total load of 100N in the X direction at the right edge. In the **Home** ribbon, **Constraints** group, select

> *Displacement*

Call the constraint **[fixed_edge]**. It will be a member of **ConstraintSet1**. Under **References**, select *Edges/Curves*, then pick the left vertical edge of the (purple highlight) modeling surface. The only available constraints at the bottom of the window are translation in X and Y for plane stress problems - there is no rotational degree of freedom. Leave both translation directions fixed for this edge. When you middle click (or select *OK*), the constraint icons will appear along the edge initially in green but will change to purple when you click on the screen.

Now we will apply the horizontal load. Pick the right vertical edge and in the mini toolbar select the *Force/Moment* button. Call the load **[endload]** (member of **LoadSet1**). Select the *Advanced* button to verify that the default distribution is *Total Load, Uniform*. Enter an X component of **100**. The Z component field is grayed out. Why? Select *OK*. To have the load arrows going outward from the model, select the *Simulation Display* button on the Graphics toolbar and set **Arrow Tails Touching**.

Defining Model Properties

So far we have not specified either the plate material, or its thickness (remember the Simulate model is only 2D geometry up to here!). These involve a major variation with what we did in the previous solid model, and with what we will see a bit later with plane strain and shell models. The command picks are non-obvious for this[3]. For 2D plane stress models, we do this using the *Shell* button 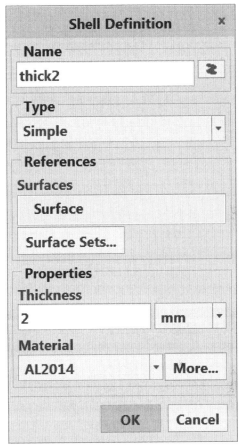 in the **Refine Model** ribbon.

This brings up the menu shown in Figure 5. Enter a shell definition name **[thick2]**. The **References** collector is active. Pick on the front (purple outline) plate surface and enter a thickness value **2**. If you hold down the RMB in the thickness data entry field, you have the option of picking a parameter for the thickness (if one has been previously defined). Beside **Material**, select the *More* button, and in the **Legacy-Materials** directory pick **AL2014** from the list. We don't have to separately assign the material as we did in the solid model of the previous lesson - shell thickness and material are both defined in this same dialog window. The completed **Shell Definition** window is shown in Figure 5. Accept the dialog and open the model tree to see the shell definition entry. Select this in the model tree (under **Idealizations**) to see the surface highlight in green.

Figure 5 The Shell Definition window to set properties in a plane stress model

IMPORTANT NOTE:

Creo does *NOT* pick up the 2D plate thickness from the 3D model (unless you use the parameter button). Recall that our actual part has a thickness of 20mm. We will see later that when we use a 3D shell idealization, Creo *will* automatically pick up the thickness from the model. **This does not happen in plane stress!**

---

[3] Notice the *Material Assignment* button in the ribbon is grayed out.

Our 2D plane stress model is now complete, and should appear as in Figure 6. Note the symbols along the constrained edges and the subtle variation in the constraint icon that indicates we are dealing with 2D stress conditions. This helps you identify which curves have been constrained.

In the Home ribbon, select the **Analyses and Studies** button, then

Info > Check Model

Hopefully, no errors are found. If there are, retrace your steps and resolve the problem.

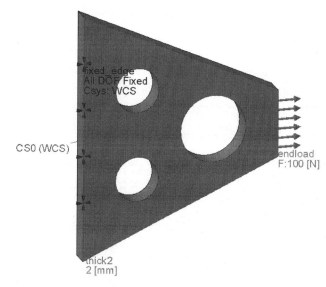

Figure 6 Completed plane stress model

Setting up and Running the Analysis

We will use the AutoGEM defaults for the mesh. In the **Analyses and Design Studies** window, select *File > New Static* and set up a **QuickCheck** analysis called **[pstress1]**. The constraint and load sets should be already selected. Check your **Run Settings** (for file directories), and then **Start Run** when you are ready. Open the **Run Status** window and scan through the run report. The run will not take very long. AutoGEM creates 19 elements (note that they are listed as 2D Plates). No errors are indicated. Make a note of the **max_disp_mag** and **max_stress_vm**.

Assuming no errors were reported with the **QuickCheck**, go back to the *Analyses* menu, and *Edit* the analysis **pstress1**. Change it to a **Multipass Adaptive** analysis with **5%** convergence on **Local Displacement, Local Strain Energy and Global RMS Stress**, maximum edge order **9** (although we hope we don't need that many). Run this analysis. The run should converge on pass 8. The final results should be maximum displacement magnitude 2.74E-03 (0.00274mm), maximum Von Mises stress 2.84E+00 (2.84MPa). Note the zero displacement in the Z direction and all stress components in the Z direction are zero as well.

Viewing the Results

Create the usual result windows for design study **pstress1**. You should be able to use the result window template for an MPA study we made previously, except for the VM stress fringe results. We are interested in the Von Mises stress fringe plot, animated deformation, and the convergence behavior of the Von Mises stress and the strain energy. When these are defined, display them.

The convergence plots for the multi-pass adaptive (MPA) analysis are shown together in Figure 7. The strain energy rises monotonically to a steady value. The Von Mises stress peaks and then comes down a bit (however note the vertical scale).

A frame from the deformation animation is shown in Figure 8. This also shows the FEA mesh. Note that AutoGEM has used a mix of triangular and quadrilateral elements here (19 total). Observe that the left edge is fixed, as desired. What is the deformation scale?

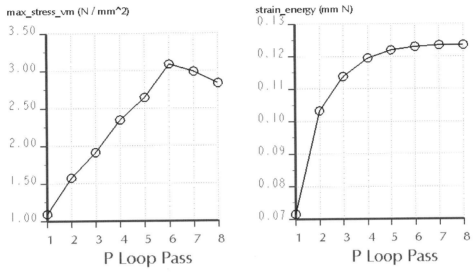

Figure 7 MPA convergence plots: maximum Von Mises stress (left) and strain energy (right)

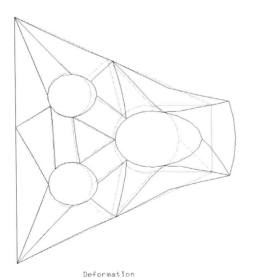

Deformation

Figure 8 Deformation of the plane stress model

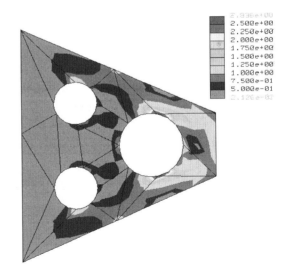

Figure 9 Von Mises stress (fringe legend modified from default)

Finally, the Von Mises stress fringe plot is shown in Figure 9. Note the location of the maximum stress on the large hole and compare this with the deformed shape in Figure 8.

Also, observe that although the mesh is not symmetric, the computed stress distribution is very nearly perfectly symmetric. The variation is due to three things: (1) we did not go all the way to "perfect" convergence (try using 1% in the MPA), (2) the coarseness of the mesh, and (3) the fringe plot is based on values of stress computed at isolated points (default 4) along each edge[4], then interpolated over the rest of the geometry.

Before you leave the results window, create a template file for these plots (for a 2D model) - you will probably want the Von Mises fringe plot, the **max_stress_vm** and **max_disp_mag** convergence graphs, and the animated deformation window. This template can then be used to quickly display results from any other plane stress (or plane strain) model. You might also have a look at a fringe plot of the element P-levels.

Note that we used constraints and loads applied along edges in this model. We saw previously that, for a solid model, surface loads and constraints were fine but point loads and edge constraints lead to difficulties with the Von Mises convergence, related to singularity in the model. For 2D geometries, edge loads and constraints are fine. However, we should anticipate problems if we used point loads and constraints in plane stress.

Exploring Symmetry

In the full plane stress model, we saw that the stress distribution and deformation were symmetric about the horizontal centerline of the model. We can exploit this symmetry in the model to simplify the FEA. In large models, exploiting symmetry can lead to very significant reductions in solution time.

Go back to the Creo Parametric interface and create a **_Thru All_** cut along the plate centerline from the left edge to the right edge. Remove the material below the cut. The model should look like Figure 10.

Transfer back into Simulate with the new symmetric half-model with

Figure 10 Symmetric half-model

> **_Applications > Simulate_**

The load and constraint are still defined. The constraint can be left alone but we will have to modify the load a bit (it is twice too big). We will also create a constraint along the symmetry boundary that is consistent with the observed (or expected) deformation.

---

[4] This setting is called the **Plotting Grid**, and is set on the OUTPUT tab in the analysis definition window.

Setting Constraints and Loads

Select the constraint on the left vertical edge. In the mini toolbar, select *Edit Definition*. This shows the **Constraint** dialog window, in particular, the constraint name **fixed_edge**, and the constraint set **ConstraintSet1**. We are going to add our symmetry constraint into this constraint set. Unlike load sets (where multiple load sets are allowed in a solution), a run can use only a single constraint set, although several can be defined for the model. So, it is important to keep your constraint sets organized and know what you are dealing with. Since we have cut away part of the constrained edge, a warning is given (small yellow dot) in the **References** collector. As long as the alternate reference (highlighted in green) is what you want, you can proceed. Close this dialog window.

For the lower (symmetric) edge select the *Displacement* button in the **Constraints** group. Name this constraint **[sym_edge]**. It is also a member of **ConstraintSet1**. Set the **References** type to *Edges/Curves* and pick both edges (using CTRL) along the bottom of the model (each side of the hole). These are the symmetry edges. The constraint to be applied here is **FREE** in the X direction and **FIXED** in the Y direction. Think about this constraint for a second - what movement does it allow of the model, and is this consistent with our expectations? The model must be allowed to stretch out along X, but we cannot allow a Y deflection since this would imply either opening a split or crack along the symmetry plane (if the edge moved up), or the plate overlapping itself (if the edge moved down). The constraints along symmetry edges (or planes in a solid model) must be consistent with the behavior of the "missing" part of the model. This can get tricky if the constraint involves rotation! Accept the dialog and observe the constraint icons.

We need to edit the load on the right end. Select the load, then in the mini toolbar select *Edit Definition*. The load is **endload** in **LoadSet1**. It should be a **Total Uniform** force. Change the X component to **50**. We have to divide it in half from the full model. Accept the dialog.

Fortunately, the shell properties (i.e. the model thickness of 2mm and the material AL2014) we set before are still defined for the model. To confirm that, open the model tree and expand the **Idealizations** entry at the bottom.

The completed symmetric half-model is shown in Figure 11. Note the constraint icon along the symmetric edge. The symbol shows that translation in the X direction is free.

Go to the *Analyses and Design Studies* window and create a *New Static* analysis **[pstress_sym]**. In the constraint and load boxes, select **ConstraintSet1** and **LoadSet1**. Set up a **Quick Check**, then *OK*.

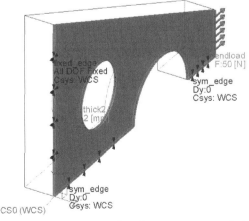

Figure 11 Symmetric plane stress model completed

Some readers might note that this model is now over-constrained. How? and Why? A necessary requirement of the constraints is to prevent rigid body motion of the plate. As long as the model is sufficiently constrained against rigid body translation (and rotation), it is solvable. Insufficient constraints will cause the run to fail. There is no problem with additional or redundant constraints, as long as they are consistent with each other and our intent. In the present model, with the constraint on Y translation on the symmetry edge, we don't actually need the same constraint on the translation of the vertical edge at the left. We have kept that constraint in this model so that we can compare results with the full plate. You might like to come back later and FREE the Y translation on the left vertical edge. This would allow a contraction (because of Poisson's ratio) in the vertical height of the plate. On the other hand, keeping the Y constraint on the left edge, and removing the Y constraint on the symmetry edge would allow the model to run, but would not be consistent with our expected deformation. Can you predict what would happen? Check that out later - you might be surprised at the result!

Running the Symmetric Half-Model

Check the **Run Settings** for the analysis **pstress_sym** and then **Start** when ready. Accept error detection. Open the **Study Status** window and look for error messages or other problems. There are only 10 elements in the half-model.

Assuming everything is satisfactory, **Edit** the analysis **pstress_sym**. Change to a **Multi-Pass Adaptive** analysis with the same settings as before (5% on default measures, maximum polynomial order 9). **Run** this analysis, deleting the existing output files. The run will converge on the 8th pass (same as before for the full model), but the run time is significantly less. The results listed near the bottom show a maximum displacement magnitude of 2.74E-03 (0.00274mm) and a maximum Von Mises stress of 2.91E+00 (2.91 MPa). These are essentially the same as before.

Create the result windows for Von Mises stress (fringe), deformation (animation), and convergence of Von Mises stress and strain energy. The convergence plots are shown in Figure 12. Compare these to Figure 7 for the previous model.

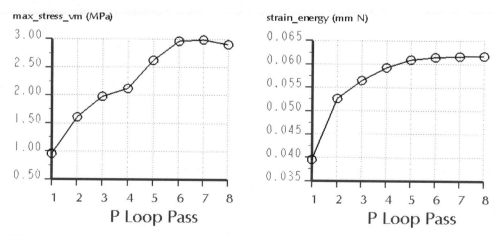

Figure 12 Convergence of Von Mises stress (left) and strain energy (right) for symmetric model

The deformation of the symmetric model is shown in Figure 13. This has the same general shape as the previous model. Observe the deformation along the lower (symmetric) edge. As required, this deflection is only in the horizontal (X) direction. The maximum displacement magnitude is identical to the full model.

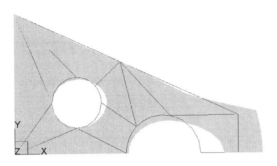

Figure 13 Deformation of symmetric model

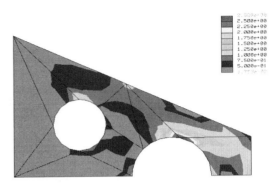

Figure 14 Von Mises stress in symmetric model (modified fringe legend)

The Von Mises stress fringe plot is in Figure 14. Compare this with the upper half of Figure 9. Notice that the fringe legend values have been manually set to have the same values. The results are qualitatively the same. The maximum value reported is somewhat different (about 2%) because of where the analysis stopped with our rather loose convergence criterion. If you rerun both analyses with a tighter criterion (like 1%), the results should be closer together. What problem might occur with the MPA for this model if you use a convergence criterion this small? (Hint: what edge order was required for convergence?) How would you resolve this?

The moral of the story here is that if symmetry exists in the model then we should exploit it to reduce model size. Symmetry can be used in all model types (solids, shells, beams, and so on). In a complicated model (or in a sensitivity study or optimization) this can cut minutes, even hours, off our execution time. Be aware that symmetry involves all of the following:

◆   symmetric geometry
◆   symmetric loading
◆   symmetric constraints (if not on symmetry plane or edge)
◆   symmetric materials (if different materials exist in the model, the same arrangement must exist on opposite sides of the symmetry plane or edge)

For example, if the **endload** we used on the full plane stress model contained a Y component, then we could not have used symmetry. In the next exercise, we will exploit symmetry even more - there are two planes of symmetry, and we only need to analyze one-quarter of the full model geometry.

There is one more type of symmetry that will be treated later in these lessons. This involves *cyclic symmetry*, which occurs when a 3D geometry is repeated numerous times in a pattern around a central axis (example: the vanes in a pump impeller or fan).

You can leave the plane stress model and return to Creo Parametric. Save the model, then remove it from the session. It might be time to take a break, too!

Plane Strain Models

A plane strain problem is one in which the geometry is defined by a 2D shape in the XY plane and the strain normal to this is assumed to be zero. Such a case typically occurs in long thin objects with a constant cross section, fixed total length, and purely transverse applied loads. Note that although the strain (ε_{ZZ}) is zero, the normal stress in the Z direction is not. This is obviously the opposite situation of the previous plane stress model.

The Model

This example will be used to illustrate a number of capabilities and functions in Simulate that we haven't used before. These include using different materials in the same model (i.e. that have very different material properties), applying a temperature load, and using some new simulation features. In order to set up this example, a number of practical (i.e. "Real Life") concerns have been neglected[5]. The scenario is the design of a long heat exchanger tube that consists of an inner core like a pipe made of magnesium alloy, and an outer jacket with longitudinal fins made of aluminum. These materials have significantly different elastic properties and coefficients of thermal expansion. This analysis will not include the solution of the heat transfer problem but is aimed only at finding the stresses due to the internal pressure and the thermal expansion of the materials. The inner pipe is pressurized to 500 psi and the entire model is elevated to a temperature 100°F above the reference temperature. Note the different units. Assume the ends of the tube are fixed so that it cannot expand longitudinally (this is what makes it a plane strain problem). The geometry of the model is shown in Figures 15 and 16.

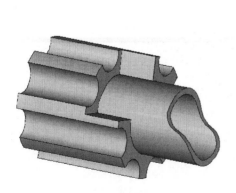

Figure 15 Cutaway shaded view of segment of heat exchanger

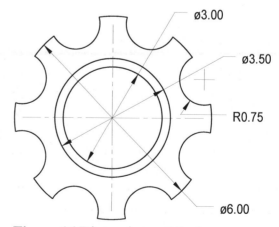

Figure 16 Dimensions of the heat exchanger cross section (inches)

---

[5] In the words of a colleague, this is a solution in search of a problem!

Creating the Creo Part

Our first task is to create the model geometry. This will be simplified since we can use symmetry about the horizontal and vertical planes in the cross section. Create a new part file **pstrain1** using the part template **inlbs_part_solid**. The part template has a default coordinate system which we will need. Recall that for 2D models, all the model entities must lie in the XY plane of a reference coordinate system. You can create just the top-right quadrant of this model (as shown in Figure 16) using a single protrusion (blind depth of 5 inches). Use **FRONT** as the sketching plane and extrude the sketch off the back side of the datum (back into the screen). Note that although the inner and outer regions are different materials, we will make a single solid to define the geometry. When we get into Simulate, we will extract the 2D shape of the cross section, then create a simulation feature to split the 2D geometry into two regions. Furthermore, the depth of this protrusion is not critical, since all we will be using is the cross sectional shape. See Figure 17 for a view of the solid model.

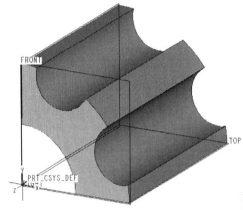

Figure 17 Symmetric quarter-model

We need to make sure that your units are set to in-pound-sec (IPS). This is not the Creo default (in-lbm-sec). Remember that Creo is very fussy about units! Transfer into Simulate and in the *Setup* dropdown, select *Units*. Pick the IPS units, then *Set* and pick the default conversion. Now we can specify the model type with

Model Setup > Advanced > 2D Plane Strain

Pick the coordinate system (containing the desired XY plane) and the front surface of the model (in the XY plane) as before. It highlights in green. Middle click. Once again you will get a warning window about changing the model type. Select *Confirm*. The edge of the front surface should now be highlighted in purple. Curiously, there is nothing in the model tree to indicate this idealization in the model, unlike the **Shell** entry for the previous plane stress model.

Creating Surface Regions

We need to divide the geometry into two regions for the different materials. We do this by defining *surface regions*. One way to create such a region is to use a curve to define a new boundary on an existing surface. We need to make the Ø3.5 curve in Figure 16. There are two ways to do this. We will use a procedure that shows the curve permanently in the model tree. In the **Refine Model** ribbon, select the *Sketch* tool. The sketch plane is the front face of the part. Use the **TOP** and **RIGHT** datums for the sketching references. Create a circular arc as shown in Figure 18. When accepted, this will appear as a blue curve on the model. Observe where this is listed in the model tree.

In the **Refine Model** menu, select

Surface Region

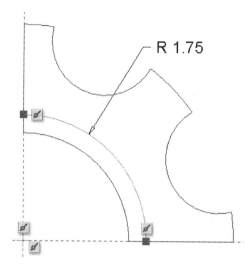

Check out the dashboard at the top of the screen.
Note that we could have launched Sketcher from
here. Follow the prompts in the message area.
First, pick on the sketched datum curve. Then
click on the inner portion where the magnesium
core will be. The inner surface will highlight in
yellow. Middle click to accept. Now if you click
on the inner surface region it highlights in green.
You will find the surface region and datum curve
in the model tree. If you click on the model tree
entry, the region will highlight on the model. If

Figure 18 Datum curve

you pass the mouse cursor over these two regions, they will individually highlight.
Notice that we don't have to explicitly define the outer region - that has happened
automatically - it is whatever is left of the 2D geometry after the first region was created.

Creating a Cylindrical Coordinate System

For what we are going to do below with the constraints, it will be helpful to switch from
the default Cartesian coordinate system to a cylindrical one. To create a new coordinate
system use the button

Refine Model { Datum } : Coordinate System

In the new window, select *Type(Cylindrical)*. Pick on the existing part coordinate
system. Accept all the defaults for defining the new system, then *OK*.

The new coordinate system (**CS0**) might be somewhat obscured. Note the shape, color,
and name of the new coordinate system icon. Go to the model tree, and select it. The Z
axis coincides with the Z axis of the default system. The X-axis label appears as "T = 0"
(for θ = 0) and the Y-axis label as "T = 90" (for θ = 90). This is consistent with a right-
handed coordinate system. We want to make this our current coordinate system, so go to
the **Home** ribbon and in the **Setup** pull-down menu select:

Current Coordinate System > Select

and pick on *CS0*, either on the screen or in the model tree. The cylindrical system now
highlights in green to show it is the current one. If you select the *Simulation Display*
command from the **Graphics** toolbar, you can turn on the option **Display Current Csys
Triad** to create a small box in the bottom right corner of the screen that shows you the
type and orientation of the current system.

Applying the Constraints

We need to apply constraints to the horizontal and vertical symmetry edges. Think about how these should be constrained for a moment. The horizontal edge cannot move vertically, and the vertical edge cannot move horizontally. We could obviously use a symmetry constraint as we did for the plane stress case, requiring two constraints. Or, combining these two and using our new cylindrical coordinate system, both edges can move radially (in the R direction) but not tangentially (in the θ direction). Since both edges are constrained in the same way, we can put both edges into the same constraint. So, in the Home ribbon select

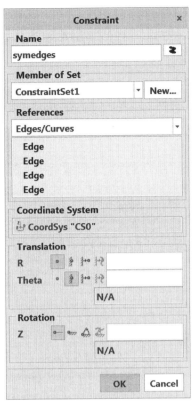

Figure 19 Creating the symmetric edges constraints

Displacement
References(Edges/Curves)

Use a constraint name *[symedges]*. With the **References** collector active, pick (using CTRL) on the two horizontal edges (one for each region) on the lower front of the part, and the two vertical edges on the left side. The picked edges turn green. Note that the coordinate system referenced is CS0 (our current system) and the constraints are in terms of R and Theta since Creo knows our current system is cylindrical. We need to **FREE** the R translation, but the Theta translation remains **FIXED**. (Leave the Rotation constraint alone. It has no effect and will be ignored if you try to set it.) See Figure 19. Middle click or select *OK*. Observe carefully the icon display for the constraints, indicating that Theta is fixed on both sets of edges.

Applying a Pressure Load

Now we'll set up the loads on the model. So that we can study the pressure and thermal loads separately, we will use two load sets.

In the **Home** ribbon select the *Pressure* button. Select *New* and call the load set **[pressure]**. In the definition window, call the load **[pres500]**. Use the collector under **References**

Figure 20 Creating a pressure load

and pick on the inside curve of the model[6]. Finally, enter a value of **500** and note that
again a RMB option here is to specify the pressure using a Creo parameter. Click the
Advanced button and note that this is a **Uniform** pressure in units of psi (lbf/in^2). Also,
for future reference note that you can apply a pressure that is a function of coordinates;
for example, you could set up a hydrostatic pressure load where the pressure magnitude
was a function of the distance down from a reference. The completed window is shown in
Figure 20. Preview the load, then select *OK*. The pressure load will show in yellow.

Applying a Temperature Load

For something a little different, we will apply a temperature load to the model. We will
apply a uniform ("global") temperature to the entire model. Using Thermal Mode, it is
possible to apply various heat loads to the part and compute a temperature distribution in
the model, then apply this temperature distribution to the model in Structure Mode[7]. We
will do this in a later lesson. By itself, in Structure, the temperature we specify is the
difference in global (the entire model) temperature from a reference temperature
(nominally 0), for which we assume the model is unstressed.

In the **Home** ribbon select

> ***Temperature***

Create a *New* load set called **[thermal]**. Enter a
load name **[tempload]**, model temperature of
100. Since we are in the IPS unit system, this
temperature is in °F. This sets the change in
temperature from the rest state (reference
temperature of 0). See Figure 21. Leave the
References selector to the default
(**Components**), and note under **Spatial
Variation** that we could set up temperature as a
function of coordinates or external field. Select
OK. A small gold symbol appears on the model
to indicate that a temperature load has been
applied (unless you have turned off the display of
load/constraint icons and values).

In the model tree, you can delete the empty
loadset **Loadset1**.

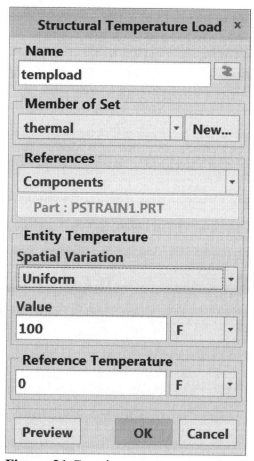

Figure 21 Creating a temperature load

---

[6] If this was a solid model, the default reference type would be a surface.

[7] It is also possible to define material properties that are functions of temperature.

Specifying Materials

Here is where we see how our surface regions are utilized. In the **Home** ribbon, select

Materials

Move the materials **AL2014** and **MG** (a magnesium alloy) from the **Legacy-Materials** library into the model by double clicking, then press OK.

Now select ***Material Assignment***. In the dialog window, the material **AL2014** is probably already picked for you. Pick on the outer region of the model (on the surface outside the datum curve). It highlights in green. You can change the assigned material name at the top of the window. Middle click. Repeat this process to select the **MG** material in the model list, then assign it to the inner surface.

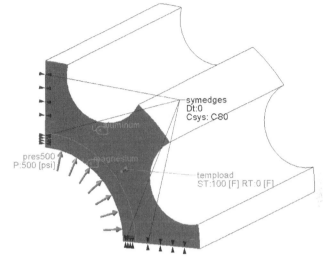

This completes the creation of the model. It should appear as in Figure 22 (the surface is shaded for

Figure 22 Completed plane strain model

clarity). Go to the ***Analyses and Design Studies*** window and select ***Info > Check Model***. If everything has been done properly, there should be no errors at this time.

Running the Model

Quick Check Analysis

As usual, we do a Quick Check to make sure that the stated problem is solvable. Create a ***New Static*** analysis called **[pstrain1]**. Enter a description. Select the constraint set **ConstraintSet1** and the load sets **pressure** and **thermal**. Select a **QuickCheck** convergence. Now you can pick

Run Settings

Check the directories for output and temporary files and RAM allocation. *Accept* the settings and then

Start Run

Accept error detection as usual. Open the ***Study Status*** window and look for error/warning messages. Note that AutoGEM creates just 5 (!) elements, identified as 2D

Solids. There do not appear to be any problems. The results show the maximum Von Mises stress for the pressure and thermal loads are 1,320 psi and 14,000 psi, respectively. Clearly, the thermal load dominates this model.

To make sure the constraints are set up properly, use **Review Results** to create windows showing the deformation animation and the Von Mises stress. Use the combined load sets in the design study **pstrain1**. Include both sets with Scaling set to 1.

Show the deformation animation (Figure 23) and note the agreement with the applied boundary conditions on the horizontal and vertical edge.

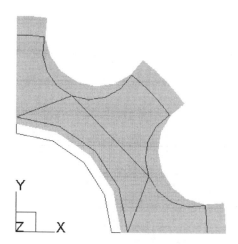

In the Von Mises stress fringe plot, we would expect symmetry about a 45° degree line. Note that the display shows line segments around the curved arcs. These connect the plotting grid points that are interpolated along each edge. The actual model does use curved arcs. Close the result windows.

Figure 23 Deformation of plane strain model (default mesh)

You might be concerned about the low number of elements. To review material from the previous lesson, let's increase the number of elements in the model to see what effect this has. In the **Refine Model** ribbon, select

　　　　　AutoGEM
　　　　　Settings
　　　　　Limits

and change the maximum allowed edge turn to 30. This should cause at least three elements along the interior arc of the model and six along the scallops on the outer edge. Accept the remaining settings.

Edit the analysis **pstrain1.** Select the *Output* tab and change the **Plotting Grid** to 8. Changing this setting should be done with caution since it affects the amount of data that must be generated, stored, and manipulated during post-processing. In the **Run Settings** dialog, check the box *Override Defaults*, and check *Create Elements during Run*. This will make sure the new AutoGEM settings are used.

Re-run the Quick Check analysis. This time there are 14 elements. No problems are indicated. The maximum Von Mises stresses for the load sets are around 14,000 psi (thermal load) and 1,500 psi (pressure load). These have changed only slightly, indicating that the default mesh with 3rd order elements was a pretty good solution.

Multi-Pass Adaptive Analysis

Assuming there were no errors in the QuickCheck, in the **Analyses and Studies** window, highlight the study **pstrain1**, then in the RMB pop-up menu select *Edit*. Change to a *Multi-Pass Adaptive* analysis with *5%* convergence with a polynomial order maximum *6*. Leave this menu and start the run.

The analysis run converges in 5 passes with a maximum edge order 5. Note the execution time - only a few seconds. The maximum Von Mises stress is about 1590 psi for the pressure load and 14,000 psi for the thermal load. These are very close to the results for the QuickCheck run above. This deserves some thought and investigation of the convergence graphs (left up to you to do!).

Viewing the Results

Since the model mesh has changed, you will have to recreate the result windows to show the Von Mises stress and deformation. Create separate windows for the pressure and temperature loads, and for a combined load with scale factors of 1 on each load set. You can make good use of the *Copy* and *Edit* commands here when setting up the result windows, or you can use the result window template for 2D models we made earlier. If you use the template, since there are two load sets, you can get Creo to produce result windows for either/both load sets, or the combined loading - a great time saver. Start with the combined loads. For the Von Mises fringe plots, check the **Continuous Tone** option.

First, display the Von Mises stress for the combined loads. See Figure 24. Notice the changed mesh and affect of the plotting grid. Then, notice that the stress is very uniform in the inner magnesium and also throughout the aluminum although it shows a slight concentration on the innermost part of the fin arcs (where the maximum stress is located). Be careful interpreting these stresses - observe the minimum value indicated on the fringe legend (it is far from 0!).

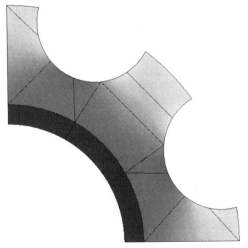

Figure 24 Von Mises stress (combined loads)

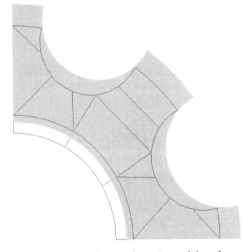

Figure 25 Deformation (combined loads)

The deformation for the combined load is shown in Figure 25. This, once again, confirms that our boundary constraints are being obeyed.

The Von Mises stresses for the separate load sets are shown in Figure 26. Note that the legend scales are different. The stresses due to the internal pressure are an order of magnitude smaller than those due to the temperature.

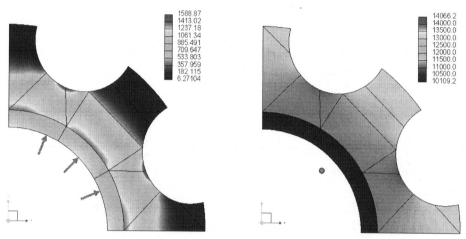

Figure 26 Von Mises stress for pressure load (left) and temperature load (right). Note that the legend scales are quite different.

You might also like to create fringe plots of the normal stress, σ_{ZZ} for both combined and individual loads. This stress is produced since we have essentially fixed the tube so that it cannot expand in length (by assuming plane strain) due to thermal expansion. This puts the tube in axial compression and is the major contributor to the high Von Mises stress related to the thermal load. Furthermore, the coefficient of thermal expansion of magnesium is greater than aluminum, so much of the load felt by the aluminum is coming from the expanding core of magnesium.

Have a look at the fringe plots for the other normal stresses, σ_{XX} and σ_{YY}. See if you can figure out the physical reason behind these stress values and patterns. In the results setup window, if you select **Component** and **XX**, you can then specify the reference coordinate system CS0 (the cylindrical system) and examine the radial (**RR**) and tangential (**TT**) stress components. How should these behave across the interface between the inner and outer regions?

Summary

In this lesson we have introduced two of the possible model idealizations available in Creo Simulate: plane stress and plane strain. The main advantage for 2D models, of course, is the speed with which the solution can be obtained versus treating the model as a solid. This is very useful for sensitivity studies and optimization, which are carried out in exactly the same way as previously presented (see the exercises at the end of the lesson). Some restrictions apply for the use of these idealizations, chiefly that the model must lie in the XY (Z = 0) plane of a chosen (or specially created) coordinate system.

Although mentioned briefly here, the problems that can arise from the use of point loads and constraints were not explored. You might do that on your own, and also see how excluding elements works on these 2D models.

We also saw how to set up simulation features (datum curves and surface regions), how to apply different material properties in the model, and how to apply a global temperature load.

In the next lesson, we will look at some more idealizations that utilize a 2D geometry to represent a 3D problem. These are axisymmetric solids and shells.

Questions for Review

Some of the following questions will require you to explore the software on your own.

1. What are the two element shapes used in plane stress and plane strain analysis?
2. How do you specify the thickness of a plate for plane stress analysis?
3. How do you assign different material properties to different regions of the same part?
4. For what conditions (what type of model) should "plane strain" be specified/used?
5. For what conditions (what type of model) should "plane stress" be specified/used?
6. What symmetry constraints (if any) would you use for the following models:

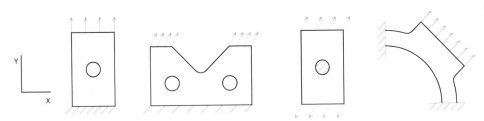

7. What are the advantages of exploiting symmetry in a model?
8. How do you create a coordinate system in Creo Simulate? Does this differ from how this is done in Creo Parametric?
9. Datum points and curves and coordinate systems created in Creo are listed in the model tree under _____.
10. Symmetry in a model includes what entities/features?
11. What are the default element limits for 2D Plates used by AutoGEM?
12. Can a thermal load use a temperature less than the reference temperature?
13. Can a thermal load involve different temperatures in different parts of the model?
14. Can you specify different thicknesses for different regions of a thin plate modeled using plane stress?

15.   What happens if you try to specify non-planar surfaces when setting the model type for plane stress or plane strain models? What happens if the surfaces are not co-planar (parallel, but not in the same plane)?

16.   Can you apply moment loads in a plane stress model? In a plane strain model?

17.   What types of constraints and loads can be expected to cause problems in plane stress models? What is the source of this problem? What is the solution?

18.   Does the use of point constraints or loads automatically invalidate results throughout the model?

19.   What is the difference between how materials are assigned to plane stress and plane strain models?

20.   What are the values of the stress σ_{ZZ} and strain ε_{ZZ} in plane stress and plane strain models?

21.   When using multiple load sets containing a temperature load, what does the scale factor mean when creating a result window? That is, what is the relation between the scale factor, the assigned global temperature, and the reference temperature? Can the scale factor be negative?

22.   When expressed in a Cartesian system, what degrees of freedom must be constrained for plane stress and plane strain models?

23.   What does the plotting grid option do?

24.   What is the meaning of the following buttons:

a)   b)   c)   d)

Exercises

1. Create a plane stress model of the thin plate shown in the figure (dimensions in mm). The plate is 2mm thick. This is the "classic" hole-in-a-plate problem. Follow the procedures as outlined in this lesson, and produce plots of Von Mises stress, deformation, and convergence behavior. Set up two models: one of the complete plate, and a second that uses symmetry. You can compare the results of the two models to each other and to analytically obtained solutions, or to tabulated values for the stress concentration around the hole (normally expressed in terms of axial stress, not Von Mises). What effect does the length of the plate have? An analytical solution will likely assume an infinitely long plate.

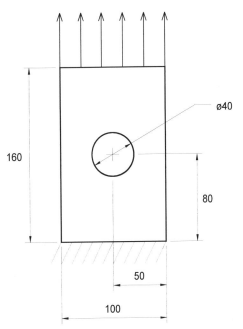

2. Perform a plane stress FEA of the following geometry and report the stress results, maximum deflection, and convergence data. The material is AL2014. Note that there are two thickness regions of the thin plate. Also, some of the plate dimensions are to the virtual sharps of the R100 fillets.

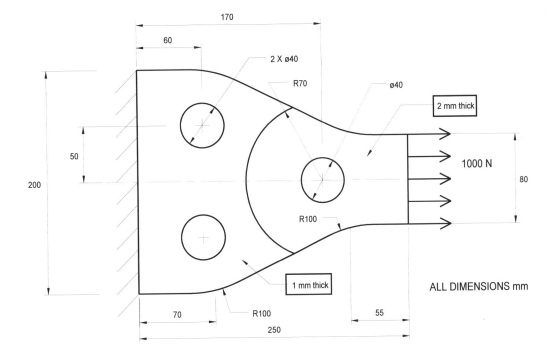

3.   Optimize the geometry of the plane stress model used in this lesson. Use the
      symmetric half model and select the design parameters shown in the figure below.
      The objective is to minimize the mass of the plate without exceeding a Von Mises
      stress of 2.5 MPa. The ranges for the design parameters are

	Min	Max
large hole diameter (**big_diam**)	30	60
small hole diameter (**small_diam**)	30	60
small hole location (**hole_X**)	40	80
small hole location (**hole_Y**)	40	60

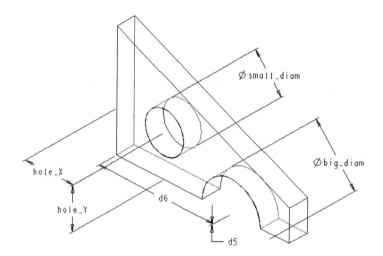

4.   Optimize the geometry of the plane strain model used in this lesson (the heat
      exchanger). The design parameters are the radius of the datum curve (at the
      interface between the inner core and outer fins), and the radius of the fin cutout.
      The objective is to minimize the mass of the model without exceeding a combined
      Von Mises stress of 14,000 psi. (Do you need to put the pressure and thermal loads
      into the same load set for optimization?) The ranges for the design parameters are

	Min	Max	Initial
datum curve radius	1.55	1.85	1.75
cutout radius	0.4	1.0	0.75

Show the convergence plots for the stress and total mass. Predict what will happen
if you increase the allowable stress to 18,000 psi, then repeat the optimization with
that value.

5.   A plane strain problem using superposition, with an interesting result.

The figure at the right shows the
cross section of a hypothetical 100'
high concrete dam. Dimensions are
in inches. The two rounded corners
are R300 (dimensions are to the
virtual corners). The dam is loaded
by hydrostatic pressure on the left
face. Assume the top of the water
reservoir is even with the top of the
dam. For the concrete (assumed
homogeneous and isotropic),
assume E = 3,500,000 psi, $\nu = 0.3$,
and $\rho = 6$ slug/ft^3. Assume the
dam is very long (into figure) so
that you can use a plane strain

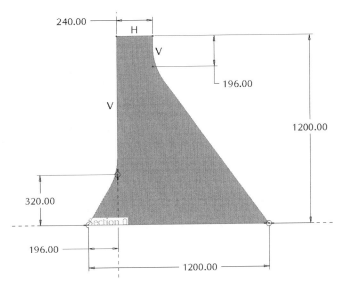

model. Using superposition (treat dam self weight due to gravity and pressure load as
separate load sets), find the following:

a) location and magnitude of max VM stress due to each load separately and the
combination,
b) fringe plots of the principal stresses in the plane of the dam caused by each load, and
c) deformation due to each load individually, and the combined load.

HINT: consider carefully how to set up the mesh and whether you will need special mesh
controls and/or excluded elements. Remember how the Von Mises stress is calculated
from the principal stresses.

(This exercise adapted from the book <u>A First Course in the Finite Element Method</u>, 4th
Ed., Daryl Logan, Thomson, 2007.)

6.   A problem showing how an FEA analysis might fit into a design problem.

The figure shows a mechanical overload protection device consisting of two plates ABE
and ACF that are free to rotate around a 10mm shaft at A. A shear pin at S prevents the
rotation caused by the load T. However, when the load *T* is large enough, the shear pin *S*
fails. The pin is 5.0mm in diameter and is loaded in double shear. The pin is made from a
suitable soft material such that it fails when the direct average shear stress is 70 MPa.
This problem concerns only the upper part *ABE*. This plate has a uniform thickness
of 3 mm so you can assume plane stress conditions. The part is made of 6061-T6
aluminum alloy for which S_y = 275 MPa. NOTE: estimate/scale unknown dimensions as
needed from the figure. **You may have to make additional and reasonable
assumptions in order to solve this problem. Although not very clear in the figure,
the designers have avoided any sharp re-entrant corners that would act as stress
raisers by using R10 fillets.**

a)   What is the load T when the shear pin fails? (HINT: This is a statics problem, not FEA! Find the load required to shear the pin at S, then do an FBD of plate ABE.)

b)   Determine the location and magnitude of the maximum von Mises stress in the upper part *ABE* when the pin fails. Think carefully about how the plate should be constrained and how the load from (a) should be incorporated.

c)   Is the thickness of part *ABE* sufficient to prevent failure based on the maximum distortion energy theory? If so, what is the factor of safety?   If not, suggest a better thickness.

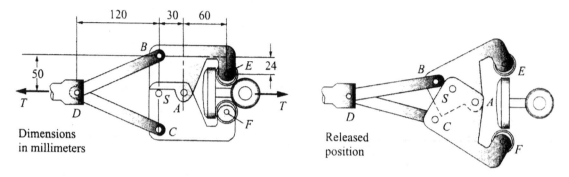

(This exercise and figure adapted from the book <u>A First Course in the Finite Element Method</u>, 4th Ed., Daryl Logan, Thomson, 2007.)

---

7.   A design problem (is it sensitivity or optimization?)

A thin plate with a hole off-centered is shown in the figure below.   A tensile load of 6,000 N is applied horizontally at each end. The plate thickness is 4.0 mm.   Determine the maximum possible diameter of the hole before yielding occurs (based on maximum shear stress failure theory). The material is A36 steel for which $S_y = 36,300$ psi.

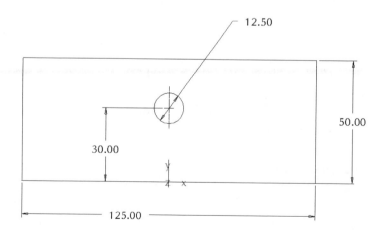

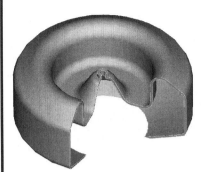

Chapter 6 :

Axisymmetric Solids and Shells

Overview of this Lesson

Axisymmetric models are another type of idealization in FEA. The objective of this chapter is to look at how an axisymmetric model is created using either solid or shell elements[1]. In an axisymmetric model, the three dimensional problem is once again represented using 2D geometry. Some new geometry commands, load types, and Simulate utilities will be introduced. A number of ways of controlling the mesh geometry will be explored that also work for 3D solid models. We will investigate pressure loads on axisymmetric shells and see how to apply a centrifugal load. An underlying message of the lesson is how an FEA model can differ from a CAD model in order to provide a more efficient solution without sacrificing accuracy. We will also look at a model whose geometry is created partly in Creo Parametric and partly in Simulate (a hybrid model).

Axisymmetric Models

Objects with axisymmetry are quite common: pipes, shafts, wheels, tanks, drums, pulleys, and rotational machinery in general. The shape of the object is defined by a planar cross-section (in Simulate this must be defined in the XY plane) which is revolved around a central axis (the Y axis). The section may or may not touch the axis, but cannot cross it. These are sometimes called revolved objects, with which users of 3D CAD programs such as Creo Parametric will be familiar. Figures 1 and 2 show cutaway views of typical examples of axisymmetric bodies. Axisymmetric models are a close cousin of models with cyclic symmetry. The difference is that the cyclic models consist of a 3D shape that appears in a regular pattern around an axis. A true axisymmetric model is defined by a 2D shape (area or curve) that is revolved to form a continuous rather than periodic geometry. We will have a look at cyclic models in a later lesson.

---

[1] Mixed models containing both axisymmetric shells and solids are also possible, but we will not have time to look at them here.

Figure 1 A solid flywheel that would be modeled with 2D Solid elements

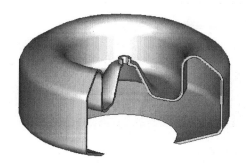

Figure 2 A hollow tank that would be modeled with 2D Shell elements

Elements

Two types of elements can be used with axisymmetric models: *2D Solid* or *2D Shell*. Both element types are defined on a single plane surface that contains the axis of symmetry. The difference between objects for these element types is illustrated in the figures above. The figure on the left would use solid elements defined on the solid cross-sectional area. The figure on the right would use shell elements that follow the cross sectional shape of the thin wall. These could be defined using a single datum curve or the edge (inner or outer) of the solid wall. For both model types, only the two dimensional shape of the cross section needs to be defined. For the axisymmetric shell, you do not even have to make a solid part - just datum curves.

Loads

A number of different loads can be applied to axisymmetric models. The major ones are:

- **Total Load** - If applied on a curve, edge, or shell represents the total load acting on the revolved surface obtained with the entity. If applied on a 2D Solid, represents the total load acting on the revolved volume. In either case, the total load stays the same when the geometry changes during sensitivity or optimization studies. Note the significant departure in Simulate from some other FEA packages, where axisymmetric loads are sometimes defined on a *per radian* of revolution basis.
- **Force per Unit Area or Volume** - If applied on a curve, edge, or shell represents the load per unit area acting on the revolved surface obtained with the entity. If applied on a 2D Solid, represents the load per unit volume acting on the revolved volume. In either case, the total load will vary if the geometry changes during sensitivity or optimization studies.
- **Centrifugal** - Load developed by a rotation around the axis of the object, specified by the rotational speed, and depends on the material mass density.
- **Pressure** - When applied to an edge, creates a force per unit area normal to the revolved surface.

Constraints

In an axisymmetric model, the axis of symmetry is fixed and is always the Y axis of the model's coordinate system. This automatically creates a constraint against free rigid body motion in the radial direction. The only other rigid body degree of freedom which must be constrained is translation in the Y direction. This will require an explicit constraint in the model. It is also often possible, and desirable if the geometry allows, to use symmetry about the radial (or X) axis.

Restrictions

When setting up an axisymmetric model, some restrictions apply. Foremost among these is that the axis of symmetry must be the Y-axis of the world coordinate system, and all model elements must lie in the right half plane $X \geq 0$.

Axisymmetric Solids

Our first axisymmetric model[2] is the thick-walled steel pressure vessel shown in cutaway in Figure 3. The inside of the tank will be pressurized. Note that the upper and lower inside corners of the tank have a fillet. It happens that the maximum Von Mises stress that occurs in this tank is at these filleted corners. It will be left as an exercise to perform a sensitivity study to examine this. We will model the cross section using 2D Solid elements. At the end of this section we will investigate some methods to control the mesh created with AutoGEM. We will also perform an analysis using 3D solid elements (that is, the Simulate default) in order to compare results and solution performance.

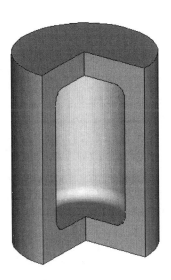

Figure 3 Thick-walled pressure vessel

Creating the Model

We can make good use of symmetry in this problem. Since we intend later to do a 3D analysis, we will set up the solid model using all three planes of symmetry (two vertical and one horizontal). In the last lesson in the book, we will revisit this model, treating it as a solid using cyclic symmetry constraints.

---

[2] Adapted from <u>The Finite Element Method in Mechanical Design</u>, Charles E. Knight, PWS-Kent, 1993, pp.165-168.

The part we are going to create is shown in Figure 4. Set up a new folder **[Chap6]** as working directory and start a new part in Creo Parametric called **[axitank]**. Change your part units to **Inch-Pound-Second** (IPS) using the *File > Prepare > Model Properties* command. While you are in this dialog window, change the part material to **STEEL**. Create the base feature as a 90° revolved protrusion off the **FRONT** datum plane. Set the revolve direction so that the protrusion comes off the back side of the datum. We want to use the face of the model lying in the **FRONT** datum (and therefore in the default XY plane) to define our axisymmetric solid geometry. The geometry of the sketch for the revolved protrusion is shown in Figure 5. Don't forget to create the axis of the revolve on the vertical reference. When the revolved protrusion is finished, create a 0.5" radius round on the inner corner. Save the part.

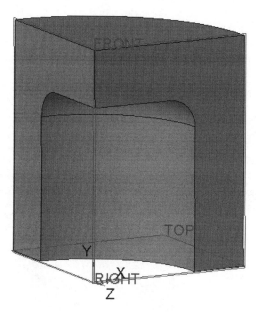

Figure 4 Symmetric 1/8 solid model

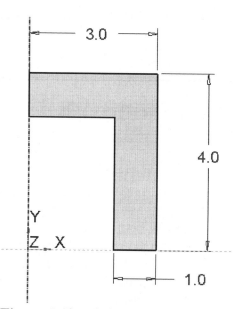

Figure 5 Sketch for the axisymmetric geometry

Setting the Model Type

When the model is ready, transfer into Simulate with

Applications > Simulate > Model Setup > Advanced

then check the *2D Axisymmetric* option. For the Coordinate System, pick the part's default system. For the Geometry reference, pick the front surface of the part. Middle click. You will have to confirm the change in model type. Turn off the datum planes. The surface is now highlighted in purple - this is the geometry for the 2D model.

Applying Constraints

Only one constraint is required, to prevent rigid body translation in the Y direction. We apply this on the lower edge of the front surface (on the X-axis). Select the command:

Displacement

Name the constraint [**midedge**]. It will be in **ConstraintSet1** by default. Set the Reference type to **Edges/Curves** and select the bottom edge along the X-axis.   Set the X translation constraint to **FREE** and leave the Y translation as **FIXED**. This allows the tank to expand radially but fixes it against rigid body motion in the Y direction. Setting a rotation constraint is irrelevant (and you will get a warning message if you try!); in previous versions of Creo, this option area was grayed out. Accept the constraint with *OK*. The constraint symbol appears on the model as in Figure 6. Use the *Simulation Display* command to control the display of the simulation entities.

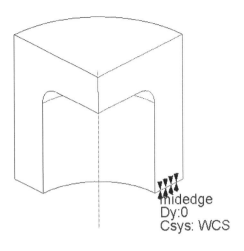

Figure 6 Constrained edge

Applying Loads

In the **Loads** group, select *Pressure*. Create a new load set called *[pressure]*, and name this first individual load *[p1000]*. Pick (using CTRL) on the three inside edges (including the fillet) of the tank. Enter the value **1000**. Note that the default under *Advanced* is *Uniform*. We could create a hydrostatic pressure field, for example, where the pressure was a function of position - useful for very tall tanks holding liquids. The completed dialog box is shown in Figure 7. Select *OK*.

In the model tree, you can delete LoadSet1 which is currently empty.

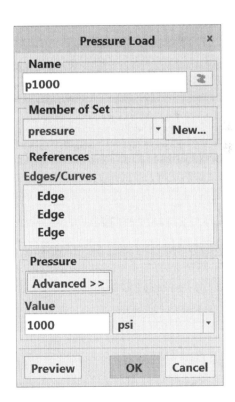

Figure 7 Creating a pressure load

Defining Material Properties

In the **Materials** group select *Materials* and
note that the material **STEEL** already appears
in the model list, since that was assigned to the
part back in Parametric. Close the window
and the 3D model outline appears in dark red.
However, observe that the material assignment
icon is still missing from the screen.
Evidently, we must still connect this material
definition to the FEA model. Select the
Material Assignment button and pick on the
model surface. Middle click or *OK*.

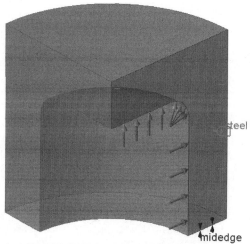

Our model is now complete. Select
Simulation Display (in the Graphics toolbar)
and under the **Settings** tab, turn **Names** on and
Values off. Set the **Distribution: Density** value to 3. The model appears as in Figure 8.

Figure 8 Model completed

Setting up and Running the Analysis

We should do a Quick Check before proceeding to see if any glaring errors are present in
the model. Starting in the **Run** group:

Analyses and Studies > File > New Static

Call the analysis name *[axitank]*. Enter a description. The constraint set is
ConstraintSet1 and load set is **pressure**. For Convergence, specify a *Quick Check*. We
are ready to run the analysis, so navigate to:

Run Settings

Select (or simply verify) the directories for output and temporary files, then

Start Run

Show the *Summary* report and review the data. There should be no error messages.
AutoGEM creates just 3 (!) elements (two quads and a triangle). Note that these are
called *2D Solids*. The maximum Von Mises stress is 3876 psi, and the maximum
deflections in the X and Y directions are 0.000181 in and 0.000217 in, respectively.
Remember that these values don't mean much with the Quick Check analysis since we
have no idea about the convergence of this data. It doesn't hurt to have a look at them,
since they will be at least the right order of magnitude. A serious blunder might show up
in an unusual value here.

Since there are no errors, we can change the analysis method to a multi-pass adaptive setting and rerun the study. Pick the analysis and using the RMB pop-up menu, select

Edit

Change the convergence to **Multi-Pass Adaptive** and set a convergence of 5% on **Local Displacement, Local Strain Energy & Global RMS Stress** with a maximum polynomial order of **9**. Then select

OK > Start Run

Delete the previous output files. Select the **Summary** again and review the run data. The run has converged on pass 8 with a maximum edge order of 8. The maximum Von Mises stress has increased to 5075 psi, and the X and Y maximum deflections have changed to 0.000184 and 0.000264 inches, respectively[3]. You might note the total CPU time required, since an interesting result will occur shortly.

Viewing the Results

Let's have a look at some results of the run. If you created a result window template for the 2D MPA study in the previous chapter, use that. If not, set up a window for the Von Mises stress (fringe plot). Under the Display Options tab, select the button to show element edges and **Continuous Tone**. Once the first result window has been set up, use **OK and Show**. Now select the **Copy** toolbar icon to create a new window *[def]* and change window parameters to produce the deformation animation (displacement magnitude, shade surfaces, overlay undeformed, animation, 12 frames, Reverse). The two windows should look like Figures 9 and 10.

Examine these results carefully. For example, check the following: is the deformation what you would expect? Are the constrained edges behaving properly? Note that the edge on the vertical Y axis stays on the axis even though there is no explicit constraint there. Do stress concentrations occur in the expected location(s)? You might like to set up result windows for the normal stresses in the X (radial), Y (axial), and Z (hoop) directions. On appropriate edges, the XX and YY normal stresses should show values of - 1000 psi on the interior surface where pressure is applied, and 0 on the outer surface. The hoop stress on the interior surface can be computed theoretically if end effects are ignored, yielding a value of 2600 psi. In the result window showing σ_{ZZ} what value do you get on the inside surface of the tank at the midplane of the model (on the X axis)? You can use a **Dynamic Query** in the fringe plots to review these values.

Before you leave the results area, have a look at the convergence graphs (Von Mises stress and strain energy). If you haven't already, save the result window template

---

[3] The results obtained by Knight (see footnote 2) using a first-order h-code analysis with 70 quad elements are maximum Von Mises = 4021 psi, max deflection X = 0.000177", max deflection Y = 0.000255"

definition by selecting *File > Save As > Type(Simulate Results Template)* and specify the name **MPA_axisolid.rwt**. We can use this template later to save some time as we explore the model mesh.

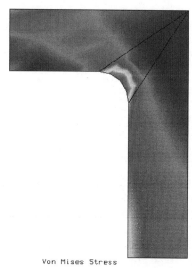

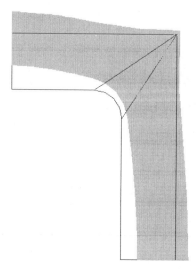

Von Mises Stress

Figure 9 Von Mises stress

Figure 10 Deformation

Exploring the Model

Changing the Mesh with AutoGEM

You might be concerned that this solution used only 3 elements. Let's investigate the effect of the mesh. Go to the **Refine Model** ribbon and select (in the pull-down menu)

AutoGEM > Settings > Limits

Change the limits for element creation to the following:

Edge Angle:	Max	150	Min	30
Aspect Ratio	Max	4		
Edge Turn	Max	30		

Accept these values. Note the warning about this value for aspect ratio - they do really mean it, although we will proceed anyway! This should increase the number of elements in the mesh, since there is less flexibility available to AutoGEM. To see the mesh that will result, select *AutoGEM > Create* and check it out. There are something like 30 mesh elements - note that tri and quad elements are color coded (light/dark green). Then go to the **Analyses & Design Studies** window and in the *Run Settings* dialog, make sure the option "Create elements during run" is turned on. *Start* the analysis and open the **Summary** report.

With the new AutoGEM settings, the run converges on pass 3 (compared to 8) with a maximum edge order of 3. The maximum X deflection is 1.83E-04 and the maximum Y deflection is 2.62E-04. These are within a few percent of the previous run. The maximum Von Mises stress is now around 5040 psi. This is pretty much the same as the previous run. An interesting result is noted with the CPU time for the run - it is very similar to the previous run (possibly even less) that had only 3 elements. This is because with so many elements (ten times as many), the MPA converges very quickly because the maximum edge order is lower. This does not happen always. As mentioned previously, your best bet is still to use the AutoGEM defaults unless you have a really good reason not to (like no convergence on 9 passes).

Create and show the result windows for the Von Mises stress and the deformation animation by opening the results template file you saved previously. These are shown in Figures 11 and 12 below. They are substantially the same as for the previous model. Increasing the number of elements with the AGEM settings has not really affected our results. Leave the results window and return to the model.

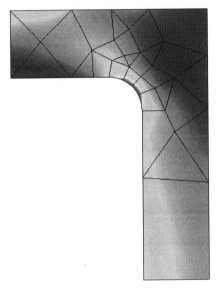

Figure 11 Von Mises stress with new AutoGEM settings

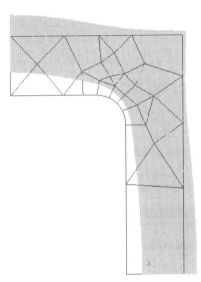

Figure 12 Deformation with new AutoGEM settings

Changing the Mesh with Detailed Fillet Modeling

As you probably know, fillets are used to reduce the stress concentration that occurs at sharp corners. Because stress gradients in the region of a fillet are usually high, it is necessary to exercise some care in setting up the model. Simulate offers a built-in option that will create a denser mesh in the vicinity of a fillet, without changing the global default AutoGEM settings on element size, edge turn, aspect ratio, and so on. This option is *Detailed Fillet Modeling*. Since we have a fillet in the current model, let's try this option to see how it works.

In the **Refine Model** ribbon, select (in the pull-down menu)

   AutoGEM > Settings

In the **Limits** tab, click the *Default* button to return all settings to their original values.

In the **Settings** tab, check the boxes beside the following options:

◆   **Insert Points** - add nodes where needed
◆   **Detailed Fillet Modeling** This is the main option we want to explore here.
◆   **Move or Delete Existing Points** While we play with the mesh, we don't want each mesh to leave remnants behind that interfere with the default action of AutoGEM.
◆   **Modify or Delete Existing Elements** We want AutoGEM to start from scratch each time we ask it to create a new mesh.

Select *OK* and then open the **Analyses and Design Studies** window, and check the settings for the study **axitank**. Make sure the box **Create Elements During Run** is checked. Go ahead and **Start** the run. Delete the old files.

Open the **Summary** window. The model contains 13 elements. The run converges on pass 5 with maximum edge order 5. The maximum X translation is 1.83E-04, and the maximum Y translation is 2.63E-04. Compare these to the previous values - pretty close! The maximum Von Mises stress is 4957 psi.

Open the **Results** window and use *File > Open* to load the results template we made previously. Show the Von Mises stress window (Figure 13) and notice the concentration of small elements created along the fillet.

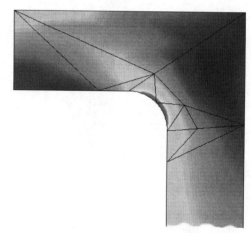

Figure 13 Von Mises fringe plot showing mesh created using detailed fillet modeling

Using detailed fillet modeling gave essentially the same results as the previous run with the much higher number of elements. This was done by concentrating elements only in the area needed, that is, at the fillet. You might investigate what happens if the fillet radius is much smaller (see Exercise #2).

Detailed fillet modeling is a global setting and produces a lot of elements in the immediate vicinity of any and all fillets in the model. It may be that only one or two of these are critical. So, how can we increase the local density of the mesh near critical fillets without changing global parameters like mesh creation limits or using detailed fillet modeling everywhere? There are a couple of ways to do this.

More Methods for Controlling the Mesh

The AutoGEM settings change grid creation parameters throughout the mesh (min and max edge angles, aspect ratio, edge turn, and so on). The detailed fillet modeling option affects fillets throughout the model. If we want to be a bit more selective about how the mesh gets modified, there are several methods available.

Go to the pull-down menu *AutoGEM > Settings*, and make sure that detailed fillet modeling is turned off, and that mesh creation limits are reset to the default values. To make sure we are getting the default three elements, use the *AutoGEM* button in the ribbon. We should get the 3-element mesh shown in Figure 9. Close the AutoGEM window and do not save the mesh.

In the AutoGEM group, select the *Control* pull-down list. Several types of control are available: distribution of nodes along edges, minimum element edge length, maximum element size, and isolation for exclusion (seen previously). We will investigate **Edge Distribution** so select that now.

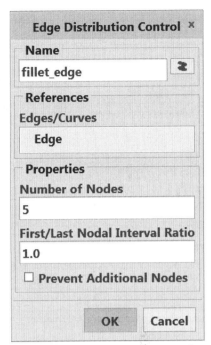

Figure 14 AutoGEM Control dialog window

Pick on the fillet edge and set the number of nodes to *5*. Enter the name of the control (that will appear in the model tree) as *fillet_edge*. Leave the nodal interval ratio setting at 1.0 (this allows you to set up a variable spacing between nodes; a value of 1.0 means equally spaced - check this out later). With the dialog completed as in Figure 14, select *OK*. The notation shown in Figure 15 appears on the model (N=5 nodes, R=1 interval ratio).

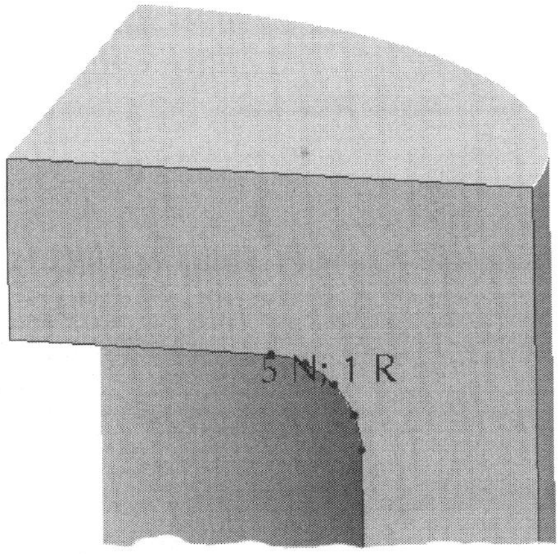

Go to the **Analyses & Design Studies** window, open the **Settings** display, and make sure the box **Create Elements During Run** is checked. The study should still be set to an MPA, with 5% convergence, maximum order 9. Run the study and open the **Study Status** listing.

Figure 15 AutoGEM control on fillet edge

AutoGEM has created 6 elements (a bit more than the default but less than with detailed fillet modeling). The run converged on pass 7 with order 7. The maximum X displacement is 0.00018 inch, the maximum Y displacement is 0.00026 inch, and the maximum Von Mises stress is about 4836 psi. These are essentially the same as before.

Use your previously saved template to display the Von Mises fringe plot, the deformed shape, and the convergence graphs. The stress fringe plot is shown in Figure 16.

Now we will check out another way that can be used to control the mesh in a specific area of the model. It is a bit more flexible than node control along the model edge, but adds additional complexity to the model. This involves creation of datum curves and nodes within the region to be meshed.

Go to the Refine Model ribbon and select the **Sketch** command. Create the curve shown in Figure 17. Sketch this on the front surface of the model. Use the **Offset** command in Sketcher to create the arc[4]. When you leave Sketcher, this will show up as a blue line. Now, create a couple of datum points along this curve as shown in Figure 17 (these are located at a ratio of 0.33 and 0.66 of the curve length, respectively).

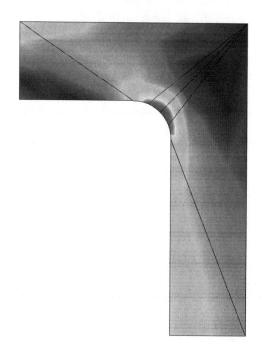

Figure 16 Von Mises stress fringe plot showing mesh created using specified nodes on fillet edge

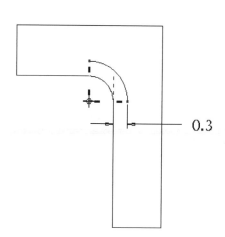

Figure 17 Datum curve to control mesh generation

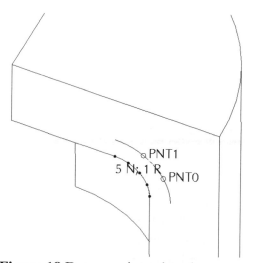

Figure 18 Datum points placed on curve to act as mesh "seeds"

---

[4] The **Offset** placement is the best choice here in case we want to change the radius of the fillet in the tank corner. A fixed radius of the curve could cause problems.

We now use the ***AutoGEM Control*** tool to specify that the curve and datum points should be used for mesh control. First, in the **Control** drop-down list select ***Hard Curve***, then create a control with the name **[mycurve]**. For the reference, select the sketched curve. Now select ***Hard Point***, specify the name as **[mypoints]**, select the **Feature** radio button, and pick either of the two datum points. Note in passing here that you can select complete patterns of points if desired. Go to the **Simulation Display** window and turn on the display of AutoGEM controls, names, and values. If you select the control in the graphics

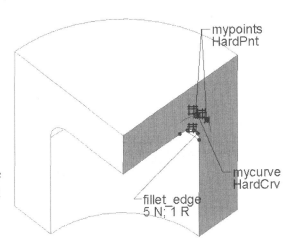

Figure 19 AutoGEM controls placed on curve and points

window, the RMB pop-up menu has a command ***Move Tag*** that lets you reposition the label to remove screen clutter. With all datums, load, and constraint labels turned off, the model should look something like Figure 19. Find these entries in the Model Tree.

Run the MPA analysis using this new control (keeping the previous edge control). The results are essentially identical to the previous results. The mesh (10 elements) is shown in Figure 20. Notice that nodes are placed at the ends of the datum curve, plus at the two datum points. Also, observe that the mesh contains both quad and triangular elements. Where are the high order elements here? Go to the AutoGEM settings, and direct AutoGEM to use triangular elements only. This results in the mesh shown in Figure 21. Results are again essentially the same as before. Where are the high order elements?

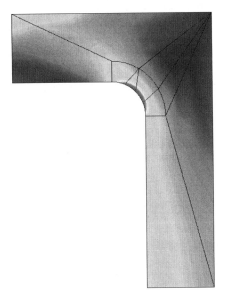

Figure 20 Mesh created using edge control plus datum curve and points (mixed mesh)

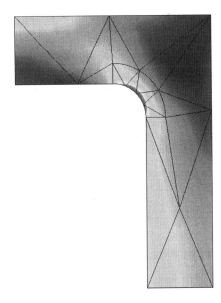

Figure 21 Mesh created using edge control plus datum curve and points (tri mesh only)

What effect has the mesh had on CPU time?

In summary, we have now seen several different methods for mesh control:

♦   AutoGEM settings - limits (edge angles, aspect ratio, etc.)
♦   Detailed fillet modeling
♦   AutoGEM Control for the number of nodes along an edge
♦   Datum curve and points for locating element nodes
♦   Selection of mesh geometry (triangles, or mixed quads and tri's)

For this simple model, these do not seem to have significantly affected the results - the variations are certainly within our 5% convergence tolerance. The variation might be reduced if we used a tighter convergence criterion but at the expense of increased CPU time. For this model, going an extra p-loop pass to obtain tighter convergence is not an issue. However, for much larger problems, the time required for the last pass can be a significant portion of the total. If that runs to many minutes, you will have to judge whether it is worth it!

These results should again give you some confidence in using the mesh creation defaults. However, you will eventually run across a problem where you will want to have more active control over the mesh geometry. A typical example would be an area of the model where stress gradients are very high, and the use of large elements means that convergence cannot be obtained even with the maximum order set to 9. It is easy to get AutoGEM to create more (and smaller) elements in just those areas. Remember also that it is possible to use excluded elements to isolate problem areas in the mesh.

It is left as an exercise to see how these ideas might extend to controlling the mesh in 3D. For example, what role do points, curves, and surfaces have in shaping the mesh produced by AutoGEM for a general 3D solid model. Meanwhile, save the model.

Comparing to a Solid Model

Before we leave this tank model, let's compare results and performance with a solid model using the full 1/8 symmetric model created in Creo Parametric. This will also give us a chance to practice creating constraints for symmetry conditions. In the **Home** ribbon, select

Model Setup

Change to a **3D** model. When you leave this menu, you are informed that all modeling entities will be deleted. ***Confirm***. Open the model tree to see the old loads and constraints are gone. We do not need the simulation features (the sketched curve or datum points), so you can delete them (right click in the model tree - it is curious you cannot suppress these). We will have to recreate our constraints and loads. In dealing with the 3D solid, we must use constraints on three faces of the model that arise from symmetry about the horizontal plane and the two vertical datums.

Before proceeding, reset the AutoGEM settings to the default values.

Now we will set up the solid model. Select

Displacement

Name the first constraint *[XYface]* in **ConstraintSet1.** Select the front face of the model (in the XY plane). For this surface, we must set the translation in the Z direction to **FIXED,** with the other degrees of freedom **FREE.** Recall that rotations have no affect on solids. Select *OK.* Continue on with the other surfaces with

Displacement

Name this constraint *[YZface]* (still in ConstraintSet1). Pick the left vertical surface. This surface must be **FIXED** against X translation with the other translations FREE.

Finally, create a third constraint on the lower surface of the solid called *[XZface].* This must be fixed against translation in the Y direction.

Come back later to check out the *Planar* constraint for these three surfaces in place of our *Displacement* constraints.

Now we can apply the pressure load. Select

Pressure

Name the load *[pressure]* in **LoadSet1**. The default reference is now a surface. Pick on the three inside surfaces of the tank (including the fillet). Enter a magnitude of **1000**.

Now assign the Material **STEEL** to the part. The completed model is shown in Figure 22. Examine the constraint symbols very carefully.

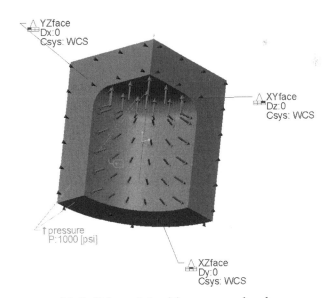

Figure 22 Solid model with pressure load

Create a new static analysis called **axitank_solid**. Run the usual Quick Check and then MPA on this model. Use a maximum order 9 and 5% convergence. AutoGEM will create 35 solid elements. For the final MPA run, results are

        convergence on pass 5 or 6
        max_disp_x   1.84E-04   (0.000184 in)

| max_disp_y | 2.62E-04 | (0.000262 in) |
| max_stress_vm | 4980 psi | |

Compare these results to the previous axisymmetric results. They are all within a few percent of each other. Notice that the total CPU time is several times as long as that for the axisymmetric models.

Create result windows to show the convergence of the Von Mises stress and strain energy (Figure 23), the Von Mises stress fringe plot (Figure 24) and the deformation animation.

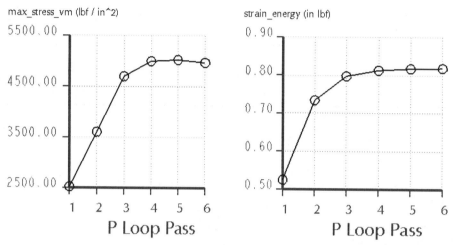

Figure 23 Convergence of Von Mises stress (left) and Total Strain Energy (right) for solid model

The convergence plots are about as perfect as we could wish. In the Von Mises fringe plot, look for axisymmetry in the fringes. Obviously we should see the same fringe pattern and values on both vertical faces. In the deformation animation, observe that the desired displacements are occurring (that is, deformation is consistent with the symmetry condition).

The thing to note here is how much CPU time was saved by using the axisymmetric model, without significantly affecting the accuracy of the results.

We will continue on, now, with another idealization that involves axisymmetric models. Leave Simulate and return to Creo Parametric.

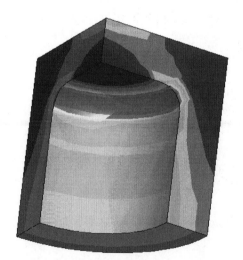

Figure 24 Von Mises stress in solid model

Remove the model from the session. This is also an appropriate place to take a break!

Axisymmetric Shells

This example will illustrate a number of Simulate functions we haven't seen before: using 2D Shell elements and applying a centrifugal load. The problem involves the analysis of the hollow axisymmetric object shown cut-away in Figure 25. We'll call it the centrifuge model. The wall thickness is very small compared to the overall dimensions, so we will use shell elements[5]. Loading on the part is due to a high speed rotation (3000 RPM) about the symmetry axis. This produces

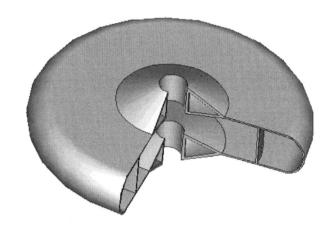

Figure 25 The centrifuge part - 400mm diameter, 3000 revolutions per minute

an acceleration at the outer rim of about 2000 g's (1g = 9.81 m/s^2). We are interested in finding out the stresses in the material, and how these are affected by the location of the vertical interior cross brace. As an alternative to the interior brace, we will also see the effect of pressurizing the interior of the model.

Creating the Model

One of the interesting things about this idealization is that no solid part is required! As for the previous axisymmetric solid, we only need a 2D representation of the revolved cross section. Furthermore, all we need to create (in Creo Parametric) is a set of datum curves that represent the walls in a cross-sectional view of the model[6]. These curves are used by Simulate to create *2D Shell* elements, which represent what would happen if the curves were revolved through 360° to form surfaces. The only thing we must be careful about is to create the defining curves in the XY plane, where the Y axis is the axis of revolution of the model. Also, due to symmetry about the horizontal plane, for this model we only need to create curves for the upper half of the centrifuge.

Start a new Creo part called **[centrifug]**. Use the **mmNs_part_solid** part template. The geometry of the model is shown in Figure 26, minus the vertical brace which we'll add later. Each thick line is a datum curve. All curves are on the **FRONT** datum. This

---

[5] In fact, an attempt to model this geometry using solid elements is probably doomed to failure.

[6] Axisymmetric shell models do not have to be created from datum curves. If we had a solid model, we could use physical edges (formed by cutting the solid on the XY plane) as the geometry references to define the model.

geometry can be created in a couple of ways:

♦   Create a pattern (using a pattern table) of datum points at the nodes and then join
        pairs of points using datum curves. This is a bit laborious!
♦   Create the curves all at once as a single sketched curve - probably the easiest way,
        and the one used here. You don't need a datum axis for the axis of symmetry -
        the Y-axis is automatically assumed.

Go ahead and create the curve shown below using **Sketch**. Create the curve on the
FRONT datum plane, and note the X and Y directions of the default coordinate system.

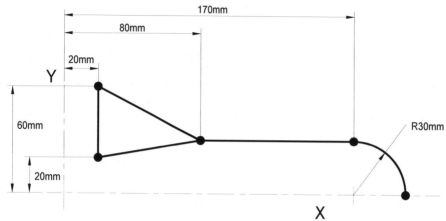

Figure 26 Dimensions for the axisymmetric half-model, without the
vertical internal brace

Setting the Model Type

When you have the sketched curve finished, go into Simulate with

Applications > Simulate > Model Setup > Advanced

Pick the option *2D Axisymmetric*. Select the default model coordinate system. Then pick
on the sketched curve - it will highlight in green. Leave the Model Setup menu with *OK*.
The curves now are highlighted in purple.

Defining Shell Properties

We need to specify the shell wall thickness and the material. We will use two different
shell thicknesses in this model. Thickness and material are both shell properties. In the
Refine Model ribbon, select

        Shell

Enter a name for this definition *[thick5]*. Note there is a button beside the name to set a color for this shell definition; the default is magenta.

Because all the edges belong to the same datum feature, we can't just pick on the desired segment - the entire feature would be selected.

Instead, put the cursor on the horizontal segment and right click. Just that segment will highlight. Now left click and the segment will highlight in bright green. To add just the arc to the selection set you must hold down CTRL when you right click again on the arc segment. Keep holding down the CTRL key as you left click on the arc. It will also highlight in bright green, and our selection set now contains both the horizontal and arc segments.

Enter a thickness of **5** (recall we are using mm). Select the *More* button and move the material **AL2014** down to the model list. Use *OK* to accept the selection. The complete shell property definition window is shown in Figure 27. Accept the definition with a middle click.

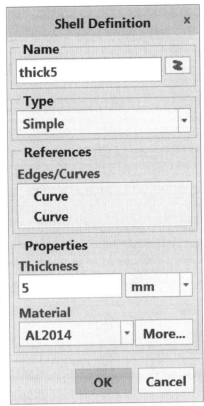

Figure 27 Shell property definition (thickness and material)

Repeat the above procedure to create another property definition *[thick8]*, and assign this property to the three curves forming the triangle at the hub. The thickness is 8 mm and the material is AL2014 as before. Open the model tree to see the shell definitions.

Setting Constraints

We need to constrain the part against moving in the Y direction only. We will constrain the vertical element at the hub (parallel to the axis) and also the point on the horizontal centerline (due to symmetry, the vertical displacement of this point must be zero). In the **Home** ribbon, select

   Displacement

Call the constraint *[endpoint]*. We will use **ConstraintSet1**. Select the **Points** option under References, and pick on the endpoint where the arc segment meets the X-axis. Set the X translation **FREE** and set the Y translation and Z rotation **FIXED**. (Why fix the rotation here?) See Figure 28. Accept the dialog with *OK*.

Now constrain the left vertical edge (the hub):

   Displacement

Name the constraint *[hub]* (also in **ConstraintSet1**). In the References list, select the **Edges/Curves** option and select the vertical edge at the left end. You will have to use a right click to preselect just the single vertical edge, then left click to accept it. This will be completely constrained (all degrees of freedom **FIXED**). Accept the dialog with a middle click or *OK*.

Setting a Centrifugal Load

To apply a centrifugal load is quite easy. The axis of rotation for 2D axisymmetric models is always the Y axis (coordinate systems other than the WCS can be selected). All we need to specify is the speed of rotation. In the **Loads** group select

Centrifugal

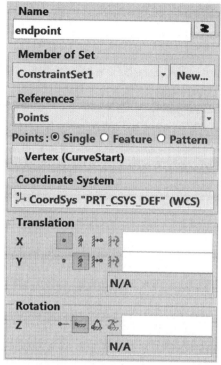

Figure 28 Constraining a point

Enter a load name *[cent]* in LoadSet1. Enter a magnitude for the angular velocity of *3000* and change the units in the pull-down list to **RPM**. The other dialog areas are grayed out at this time. For an axisymmetric model, the axis of rotation is always the Y axis (note the direction vector components < 0, 1, 0 >). For general 3D models we can specify the axis of rotation using a vector direction defined by three components points. This is not available or required for 2D models. Accept the dialog. Note the gold centrifugal load symbol on the Y axis (this is easier to see if you turn off coordinate system display).

The model is now complete. In the model tree, right click on Loadset1, and select *Rename* in the pop-up menu. Change the name to *[centrifugal]*. We will be adding a second load set for the pressure load a bit later.

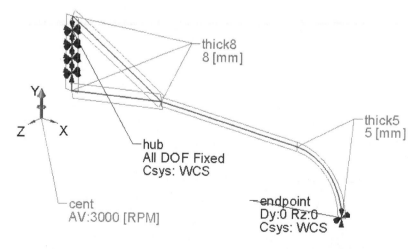

Figure 29 The completed axisymmetric model (isometric view)

Running the Analysis

As usual, the first time we run a model, we will perform a Quick Check to see if there are any serious modeling errors. Select:

Analyses and Studies > File > New Static

Enter an analysis name *[centrifug]*. Enter a description and make sure the defaults **ConstraintSet1** and **centrifugal** are highlighted. Finally, select **Quick Check** and leave the dialog.

Check the **Run Settings** and select the desired directories for output and temporary files. Then select **Start**. Check the **Summary** report for any reported errors or warnings. You will note a warning about stress error estimates - this will sometimes occur when the maximum stress occurs on a constrained entity. There are 5 *2D Shell* elements. You might note the maximum Von Mises stress is 9.4 MPa; maximum X and Y deflections are 0.014 mm and -0.019 mm, respectively.

Since there are (or should be!) no errors, we can change the analysis to a multi-pass adaptive analysis with **5%** convergence on **Local Displacement, Local Strain Energy & Global RMS Stress**, and leave the maximum polynomial order of **6**. You will find that shell models seldom need higher order elements. To get a bit more resolution in the display of the displaced shape, go to the **Output** tab and change the **Plotting Grid** setting to **10** (default 4). This will give us smoother curved shapes. Accept the dialog, then

Start Run

Delete the existing output files for the model. Open the **Summary** window. The analysis converges on pass 3 with a maximum edge order 5 (what order was the first pass?). The maximum Von Mises stress has increased to 11.2 MPa; maximum X and Y deflections have also increased to 0.0186 mm and -0.0219 mm, respectively.

View the Results

Create the usual result windows for the Von Mises stress and a deformation animation. Call the first window *[vm]*. Get the output directory **centrifug** from the location you specified under **Run Settings**, enter a window title, set **Display Type(Fringe)** and the **Quantity (Stress, Von Mises)** and plot the maximum stress value of the top/bottom surface. Select *OK and Show*. Using the ribbon button, *Copy* the window definition to a second window called **[deform]** and modify the definition to set up a deformation animation with an overlay showing the undeformed shape.

Make the von Mises stress window active and select

Model Max

A label is placed on the model to show the location and value of the maximum stress. See Figure 30. If you use **Dynamic Query** (in the RMB pop-up), you can determine stresses at other points in the model, but this is a bit tricky since you must be very accurate in picking the plotting grid points.

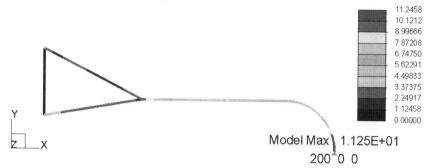

| 11.2458 |
| 10.1212 |
| 8.99666 |
| 7.87208 |
| 6.74750 |
| 5.62291 |
| 4.49833 |
| 3.37375 |
| 2.24917 |
| 1.12458 |
| 0.00000 |

Model Max 1.125E+01
200 0 0

Figure 30 Von Mises stress in axisymmetric shell model with centrifugal load

Figure 31 shows the (exaggerated) deformed shape. What is the displacement scale? Recall that the maximum Y displacement is only 0.02mm.

Figure 31 Deformation of the axisymmetric model with centrifugal load

From the deformation we see that the horizontal wall of the shell is drawn inwards. We will add a vertical brace inside the shell to stiffen it in the Y direction. Before you leave the **Results** window, have a look at the convergence graphs.

Modifying the Model

For all our models up to this point, we have created the geometry in Creo Parametric and brought it into Simulate. We are going to add to this geometry in Simulate to create what might be called a hybrid model. It is important to realize that the simulation feature we add here is not known to Parametric and will disappear when we leave Simulate[7]. In fact, it would be possible to create this entire model within Simulate using simulation features

[7] To make the simulation feature visible in Creo Simulate, it must be "promoted". See the RMB pop-up for the feature (in the model tree) after we are finished.

alone. This only applies when there are no solid features in this geometry. Those cannot be created in Simulate.

In the **Refine Model {Datum}** group, select

> *Sketch*

Pick on **FRONT** for the sketching plane, and **TOP** as the top reference. In Sketcher, pick the horizontal datum curve as a reference. Create a vertical line at X = 130 between the X-axis and the horizontal element. Make sure the top vertex is aligned with the horizontal curve. See Figure 32. Accept the sketch and observe the color of the line we just created. It is different because it is currently not in the model. Go to

> *Model Setup*

Select the collector under **Geometry**, and CTRL-click on the new curve. Middle click to accept. It is now included in the model but is not yet a shell (notice the line thickness).

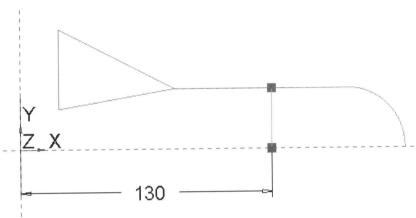

Figure 32 Sketch of simulation feature for the vertical brace

We need to apply properties to this new geometry item (we'll just add this curve to the **thick8** group). Select one of the three current edges in the **thick8** group (all three should highlight), then in the mini toolbar, select *Edit Definition*. CTRL-pick on the new datum curve. Middle click. The shell set **thick8** now contains four curves. Verify this by selecting the group in the model tree.

We need to add a symmetry constraint on the lower point on the new brace. This is easiest to do by adding the endpoint of the new curve to the existing constrained end point definition. Do that by selecting the constraint, then in the mini toolbar, select *Edit Definition*. Add the new curve end point. The new model should look like Figure 33. Open up the model tree and expand all the branches to see the data structure.

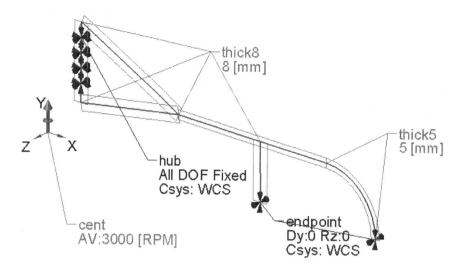

Figure 33 Axisymmetric model with additional brace (isometric view)

Running the Modified Model

Open the *Analyses and Design Studies* dialog. We shouldn't have to make any changes to the analysis type, so we can go directly to

Run Settings

and check your settings. Make sure "Create elements during run" is selected (since we have added a new shell). *Start* the analysis and open the **Study Status** window. Note that there are now 7 elements - the horizontal curve has been split at the vertex/junction with the vertical brace. The run converges in 3 passes with a maximum edge order of 5. The maximum Von Mises stress is reduced from the previous value of 11.2 down to 9.00 MPa. The deflections are now 0.0146 mm and -0.0097 mm in the X and Y directions, respectively. The Y displacement has been cut in half from the previous model.

Create the same result windows as before (easy to do if you have a 2D results rwt file!). Open the window showing the Von Mises stress and use *Info > Model Max* to have a look at where this occurs. Results should be as shown in Figure 34.

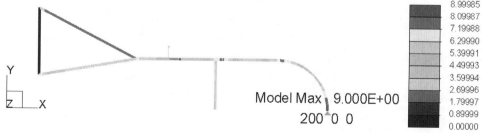

Figure 34 Von Mises stress in modified model

The deformation of the modified model is shown in Figure 35. What is the deformation scale? The vertical brace has partially prevented the collapse of the side wall, as intended, and also served to reduce the maximum stress in the model, which still occurs at the outer circumference.

Figure 35 Deformation of the modified model

It would be interesting to find out how the location of the vertical brace might affect the stress and deformation in this model. From Figure 35, it looks like moving it farther out might reduce the vertical collapse even more. This problem is left as an exercise.

Before we move on, and to prepare the model for the exercises, in the model tree under **Simulation Features** select the sketch for the vertical brace curve and in the RMB pop-up menu select *Promote*. The curve is moved out of the **Simulation Feature** entry in the model tree and becomes a feature in the Parametric part itself. Leave Simulate in order to check that out, then suppress the curve (and its children) and come back to Simulate.

Pressure Loads on Axisymmetric Shells

Another way that might reduce the sidewall collapse is to pressurize the inside of the centrifuge (or evacuate the outside).

Pick the *Pressure* button in the **Loads** group. Create a new load set called *[Pressure]*, to keep it separate from the centrifugal load. Name the load *[pres]*. The References are automatically going to be edges. We will select two edges, but they must be in separate loads for reasons we will see shortly. Select the horizontal edge of the model (use the RMB to query select). Then enter a value of **100** with units set to **kPa** - this is about 1 atmosphere. If you select *Preview*, you will see the load arrows pointing downward - this is a consequence of which way Simulate is considering the positive normal direction to the edge. We want the pressure to act upwards, so change the value to **-100** (kPa). Select *OK*. The load arrows will show in yellow.

Select the *Pressure* button again. Keep the load in the set **pressure**. Name this load something like *[end_arc]*. Select the arc on the end of the model. Enter a value of **100 kPa**. (Check to see what happens to the pressure icons if you accidentally leave this in MPa.) Select *Preview*, and enter a negative value if necessary to have the arrows pointing outward. Select *OK*.

To make sure our pressure load is going the right way, in the **Loads** group pull-down menu select

Review Total Load

Pick the **Loads** button and select the pressure load on the arc, then middle click. Then select ***Compute Load Resultant***. The resultant of this pressure load is indicated to be 3487 in the positive Y direction. Why is there no resultant force in the X direction? Can you confirm that this is the correct value? (HINT: remember the model is axisymmetric.) The important thing here is that the resultant is upwards. Perform the same check with the pressure on the horizontal curve.

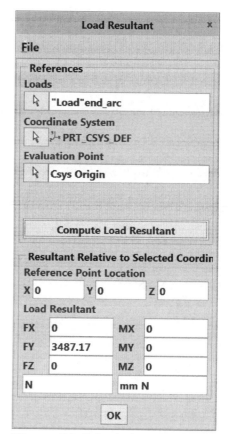

Figure 36 Calculating the Load Resultant for the pressure load on the arc

Go to the **Analyses and Design Study** window. Copy the study **centrifug** to create a new study **centrifug_total**. Edit the new study to make sure both load sets (**centrifugal** and **pressure**) are selected. Go straight to an MPA convergence analysis. Start the run, remove existing files, and open the **Study Status** window. The run will converge on pass 3 with a maximum edge order of 5. For the centrifugal load, the maximum von Mises stress is 11.2 MPA (as before). For the pressure load set, it is 26.9 MPa. Note that for the pressure load, the maximum displacements in the X and Y directions are in the opposite directions from the centrifugal load.

Create separate result windows for the deformation due to only the centrifugal and pressure loads. When you create these windows, set the scale factors for each load set to either 1 or 0 in order to isolate the two loads. Set the deformation scales in each of these windows to 200. See Figure 37. Clearly, the pressure is pushing the sidewall outward.

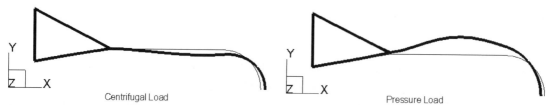

Figure 37 Deformation due to 3000 RPM centrifugal load (left) and 100 kPa pressure load (right). Scale factor 200.

Create a result window showing the deformation with the two loads acting simultaneously. Set the scale factor on the pressure load to 0.3. This puts an internal pressure of 30 kPa (about 4.5 psi) inside the centrifuge. Copy the window definition for the combined loading and edit it to show the Von Mises stress. The maximum von Mises stress is now 8.6 MPa (this is less than we had with the vertical brace).

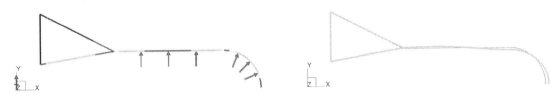

Figure 38 Von Mises stress (left) and Deformation (Scale = 200) (right) for combined loading of 3000 RPM and 30 kPa internal pressure. Maximum stress 8.6 MPa.

By keeping separate load sets for the centrifugal load and the pressure, we could conceivably find what pressure would be optimal (produce minimum stress levels) for several possible values of the centrifugal load. Here is a question for you to investigate (see Exercise #4): If we double the scale factor on the centrifugal load result window, is this the same as doubling the rotation speed of the centrifuge, or some other factor?

Summary

Axisymmetric models are quite common, and you should be familiar with both solid and shell elements. Fortunately, these are easy to set up and the computational load is very light, so the models will execute quickly. Axisymmetric models can also combine solid and shell elements (consider a flywheel with a thick inner hub and outer rim connected by a thin disk). We will see one of these combined models in the next chapter. You should try to do some of the exercises below to get more practice at creating and interpreting these models.

Questions for Review

1. On what plane do you have to create an axisymmetric model? Where is the axis of symmetry? How is this indicated in the model?
2. What types of elements are available for axisymmetric models? Make some quick sketches or describe typical shapes appropriate for each element type.
3. What are the main load types available for axisymmetric models? How are the Simulate definitions different from standard FEM package definitions?
4. What is the minimum constraint required in an axisymmetric model?
5. What are three ways you can modify the default mesh created by AutoGEM?

6.   Is it possible to have an axisymmetric model which, under load, will deform such that it crosses the Y-axis?

7.   At what point in the model creation do you identify it as a 2D axisymmetric model?

8.   At various times, sketched curves may be displayed in red, blue, magenta, or green. What is the significance of these colors?

9.   How can you apply pressure to an axisymmetric shell?

10.   What will generally happen to the convergence when you produce a finer mesh in Simulate? Why?

11.   What will happen to the mesh created by AutoGEM if you sprinkle datum points on the model?

12.   How do the usual AutoGEM limits on edge turn work on datum curves placed on the model?

13.   Is it possible to create 2D Solid elements that overlap? What about 2D Shell elements? First, do you think this is a reasonable thing to be able to do, and second, find out how Simulate will respond if you try to do this.

14.   Find out if it is possible to create shell elements of varying thickness along an individual curve segment. If not, how would you approach modeling this?

15.   What is the *lowest* edge order used in a multi-pass adaptive analysis of a 2D Shell model?

16.   When specifying a centrifugal load, what units are used for the rotational speed?

17.   What happens if you apply a gravity load in the X direction in an axisymmetric model? If there is a gravity load in the Y direction, can we still use symmetry about the horizontal plane?

18.   How are the material and thickness specified for a 2D Shell?

19.   How is the material specified for a 2D Solid?

20.   Why is no radial constraint required in an axisymmetric model?

21.   How can you create mesh nodes with variable spacing along an edge so that they are denser at one end of the edge than the other? Can you also do this for surfaces?

Exercises

1. For the axisymmetric solid tank model, what is the effect of the fillet radius on the maximum von Mises stress? Investigate this using a sensitivity study with the radius varying from 0.1 to 1.0 inch. Try this with and without detailed fillet modeling and comment on the results. In particular, what is the effect of detailed fillet modeling for large and small radius fillets?

---

2. Consider a 3" diameter circular steel rod with a circumferential semi-circular groove under an axial tension load.

(a) Assuming that the load is 100 lb, find the maximum axial stress in the rod if R=0.25".

(b) Do a sensitivity study for values of R ranging from 0.10" to 0.50" in order to find the variation in the maximum axial stress as a function of R. Plot these results in terms of a stress concentration factor, with the axial stress normalized with a nominal stress based on the minimum rod diameter. (HINT: remember you can export graph data to a spreadsheet.) Compare this factor with published values (you'll find these in mechanical design textbooks). Justify your choice for the dimension L. Comment on the use of FEA for obtaining stress concentration factors. (HINT: You will have to play around with the mesh to get reliable results.)

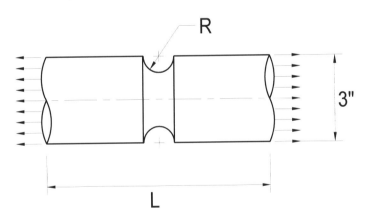

---

3. In our centrifuge problem, we want to find out the effect of the location of the interior brace. This calls for a sensitivity study. Set this up using the axisymmetric shell model with the location dimension to the curve in Simulate as the design parameter. Vary this distance from 100 to 160 mm. Plot the variation of the maximum Von Mises stress and maximum displacement magnitude with this parameter.

---

4. As suggested in the text, find the relation between the load scale factor for the centrifugal load, and the specified rotation speed in the load definition. That is, if the scale factor is doubled, what is the effective change in speed?

5.   There is a lot of interest in the use of
      flywheels as energy storage devices, in
      everything from power stations to
      automobiles and buses. One of the
      main design issues is the stress caused
      by the high rotation speeds (up to
      60,000 RPM!). A design concept for a
      flywheel is shown in cutaway at the
      right. For a given speed, the maximum
      energy storage is obtained by
      maximizing the inertia about the

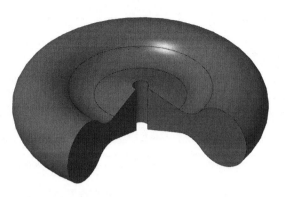

rotation axis. At the same time, the stress cannot exceed a limiting value. The
question is what set of geometric parameters will satisfy these requirements? This
calls for an optimization solution, using geometric parameters of the cross section
shown in the figure below (dimensions in millimeters).

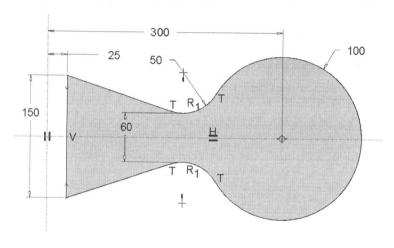

Set up an optimization design study as follows:

Objective - maximize the inertia Iyy (you can determine this using a **Measure**, defined
      in the **Home** ribbon)

Constraint - Von Mises stress not to exceed 2000 MPa

Design Variables	Min	Max
rim radius (100mm as shown)	70mm	150mm
hub height (150mm shown)	50mm	250mm
(The other dimensions in the section are not variable.)		

Assume the flywheel is made of steel and has a rotation speed of 15,000 RPM. You will
first have to set up an efficient and reliable static stress analysis, involving choice of
model, constraints, mesh control, etc. Report on the initial values of the goal and the max
VM stress (and where it occurs). Then run the optimization design study and report the
results.

6.   Make a solid model of the
centrifuge, as shown in the figure at
the right. (HINT: use the geometry
of the sketch in Figure 26, then use
Thin, Revolved Solid). Bring this
into Simulate and model it as an
axisymmetric solid. The material is
AL2014. Apply the same
centrifugal load (50 rev/second).

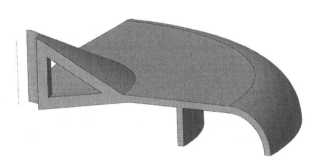

Perform a static stress analysis on the given geometry and report the convergence
behavior and the relevant stress and deformation values. Once you have a stable static
analysis, set up an optimization to find the minimum mass without exceeding a stress of
25.0 MPa. The design variables are the thickness of the shells (***thick5*** and ***thick8*** in the
previous model) and the radial location of the vertical support. How much was the mass
reduction from the initial design? Show this in a graph (mass vs optimization iteration).

---

7.   Continue the optimization of the previous question adding the following factors: The
exterior of the centrifuge is an evacuated chamber (to reduce air friction) while the inside
of the centrifuge is pressurized to 30 kPa as shown in the figure below. In addition, we
cannot allow the maximum horizontal or vertical deflection anywhere to exceed 0.02
mm. Find the thickness of the shells and the location of the vertical support that will
result in the minimum inertia around the axis of rotation. For the optimized design, which
constraints are active? What is the maximum stress and where does it occur?

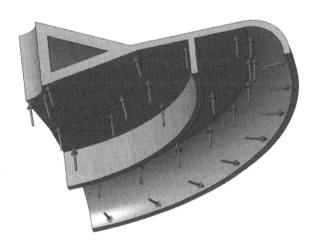

8.   A crude design model of an axisymmetric expansion fitting for a steel piping system is shown in the figure below (cutaway view). This model (**expandjoint.prt**) can be downloaded from the SDC web site. The model is loaded by an ambient temperature 100°C above reference plus an axial movement of the two flanges (caused by the expansion of the connecting pipes). The flange movement compresses the fitting by 0.5 mm in the axial direction (i.e. overall length is reduced by 0.5 mm). Although the temperature is uniform in the model, thermal expansion will produce stress in the material because the ends are constrained axially.

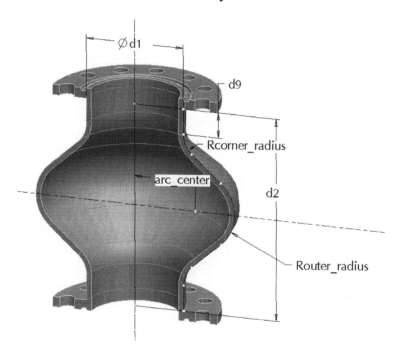

The design problem is to find out what shape the fitting should be to obtain the lowest maximum Von Mises stress in the fitting wall. This shape is determined by three dimensions in the figure. The permissible range of these parameters is:

Router_radius	min	50	max	125
Rcorner_radius	min	25	max	75
arc_center	min	100	max	250

Note that it may not be possible to regenerate the model for all possible combinations of these parameters - that is, some combinations may result in illegal geometry.

Set up an optimization design study to determine the appropriate values for these design variables. What is the value of the Von Mises stress for the initial and final designs? Hint: There are at least three (or four, if you jump ahead to the next lesson!) different model types that could be used for this problem. Pick the one you feel is most appropriate and be prepared to explain why.

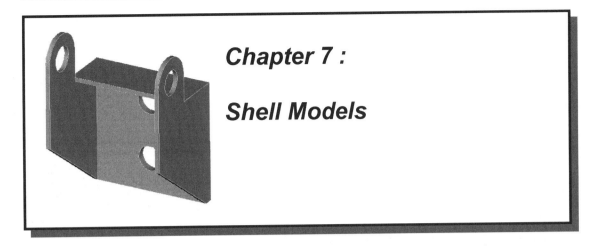

Chapter 7 :

Shell Models

Overview of this Lesson

For 3D modeling that is not axisymmetric, after the default solid elements the next most common element is the shell. In this lesson we will investigate idealizations using shell elements to represent solids defined by pairs of parallel surfaces. In the previous lesson we saw that an axisymmetric shell model could be constructed using datum curves. The shell properties (thickness and material) were specified using a dialog window and then assigned to the geometric curves. In particular, the shell thickness was supplied as input data rather than coming from the Creo model itself. The same situation occurred with the plane stress model earlier.

For general 3D shell models, things are a bit different. There are two types of shell definitions. The simplest is a shell that is defined on a model surface. In this case, as in the axisymmetric shell, the thickness is a shell property and not read from the model. The second type is the *shell pair*, which we will deal with exclusively in this lesson. In this case, Simulate reads the shell thickness directly from the Creo solid model. Shell models (or the portion of the model to be represented using shells) are determined by pairs of surfaces, called the shell top and shell bottom. The two defining surfaces in a pair are *compressed* to a mid-surface location where the shell elements will be created. In this lesson, we will deal with shells of constant thickness only. And, of course, the shell thickness(es) being a dimension in the Creo model can be used as a design parameter(s) for sensitivity studies and optimization.

The purpose of shells is to produce more efficient models. If portions of a solid model are composed of thin-walled features, treating them as solids is very inefficient (and sometimes prohibitive). The number of solid elements required to represent these features can be enormous. The general guideline for using a shell is that the thickness dimension should be less than about 1/10th of the length of the shortest edge of the shell surface. It is possible to put shell elements on tightly curved corners, but the radius should be several times the shell thickness. We will investigate this in one of the models studied in the lesson.

We will look at three simple examples to illustrate the procedures for analysis. In the first example, the surface pairs forming the shells are determined automatically. In the second example, we will identify the pairs manually. This model will also illustrate a problem (unrelated to how we made the shells) with obtaining convergence and let us explore the capability to exclude elements from the convergence analysis. In the final example, we will create a model containing both solid elements and shells.

Automatic Shell Creation (Model #1)

Creating the Geometry

We will analyze the small pressurized tank shown in Figure 1 (approximately in default orientation). The numbered features show a reasonable feature creation order. The tank has a hemispherical shape on the bottom. Start a **New Folder** *[chap7]* and create a new part called *[shelltank]* making sure that you set up the units as mm-N-s. The dimensions (in mm) of the major features of the tank are shown in Figure 2. Note that features #2 (φ15mm) and #3 (φ20mm) are sketched on TOP and RIGHT, respectively, as shown in Figure 2 and the round dimensions are given in Figure 1. Use the *Shell* feature last to create a wall of uniform thickness (1.0mm) throughout, removing the flat surfaces to the two cylindrical extrusions. To take advantage of symmetry, create a final vertical cut through the model to remove the front half (as in Figure 2). Turn off all datums (planes, axes, etc.).

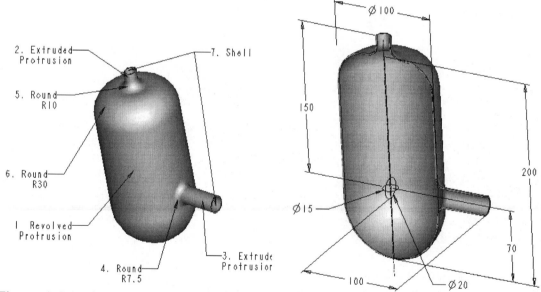

Figure 1 Creo features used to form tank

Figure 2 Tank dimensions (mm)

Defining the Shells

When the geometry is complete, select

Applications > Simulate

Recall that the default model type is 3D, so we don't need to change that.

In order to see the shell pairs created in the next step, go to the **Simulation Display** window and in the **Modeling Entities** tab make sure that the **Idealizations(Shell Pairs)** is checked (on).

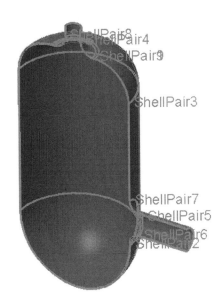

Our next job is to define the shells in the model. In the **Refine Model** ribbon in the **Shell Pair** drop down list select

Detect Shell Pairs

Turn **OFF** the option to **Use Geometry Analysis**. When this checkbox is cleared, Simulate will detect paired surfaces that are implicit in some types of features (shells, ribs, sheet metal, thin protrusions)[1]. Then select **Start**. Outer and inner surfaces will highlight in green and cyan. See Figure 3. If you **Repaint** the screen, the edges will appear in orange and magenta.

Figure 3 Automatically detected paired surfaces

Then in the AutoGEM group select

Review Geometry > Apply

This will show the midsurface that will be used to create the shell elements. In the **Simulation Geometry** window you can toggle the display of the tank's **Original Geometry**. A close-up of the top of the tank is shown in Figure 4. Note that this window would let you locate areas of unpaired and unopposed surfaces in the model, either of which would cause problems later. If, as we will do later, you create the shells manually, this will help locate any shells you might have missed.

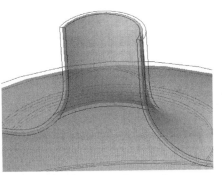

Figure 4 Closeup of top pipe showing the midsurface and original geometry

Close the **Simulation Geometry** window. Open the model tree to see the shell entries that are listed under **Idealizations** - there should be 9 shell pairs. Selecting these in the model tree will cause them to highlight (in green and cyan) on the model.

---

[1] When the box is checked, Simulate will look for all shell pairs with a thickness less than the specified characteristic thickness. These surfaces can come from different features. This option lets you control which volumes bounded by parallel surfaces will be modeled as solids and which as shells. See Model #3 and Exercise #5.

Go to the pull-down menu in the AutoGEM group. Notice that the option **Solid/Midsurface** is now checked. Since we don't have any solids in the final model, change this to **Midsurface**. The next time you create a mesh for this model, AutoGEM will automatically use the paired surfaces to create shell elements on the midsurface.

While you are in the AutoGEM menu, make sure your AutoGEM limits are set to the default.

Assigning the Material

We now need to assign the usual materials, constraints, and loads to the model. Start with the material. In the **Home** ribbon, select

> ### *Materials*

Bring the material **SS** (a stainless steel) from the Legacy library into the model and select **OK**. Then select

> ### *Material Assignment*

Check out the contents of this window, then select **OK**. Thus, material assignment is done in exactly the same way as for a solid model. Now that a material has been assigned, you could launch AutoGEM to see what kind of mesh is created. We will see that in a few minutes.

Assigning the Constraints

In 3D solid models, we normally apply constraints to surfaces. For the symmetry constraint in this model, that would be the thin surfaces created by the symmetry cut. However, those surfaces will disappear when the shell surface is created by compressing the surface pairs. So instead we will apply constraints to shell edges. We can pick either the inner or outer surface edge - both will collapse to the midsurface. **It is important to note that, unlike solid elements, shell elements have rotational degrees of freedom.** So, remember that for shell models:
1. we must apply constraints to edges or curves, and
2. we must keep rotation of those edges in mind (i.e. constrained or not).

Creo contains a constraint command (***Symmetry***) that is a shortcut for creating constraints involving either mirror or cyclic symmetry. As usual, we will forego this shortcut for now to see what is actually involved in setting up the mirror symmetry constraint for the tank. You should come back later to explore the use of this command.

In the ribbon, select

> ### *Displacement*

Call the constraint *[symedges]* (member of **ConstraintSet1**). Set the **References** type to *Edges/Curves*, and go around the outer (magenta) edge of the solid model on the symmetry (XY) plane picking all the edges of the tank using CTRL to add edges to the selection set. You may have to zoom in to do this. Each edge will highlight in green when selected. There are 17 edges forming two tangent chains - note the counter beside the selection filter at the bottom[2]. Symmetry requires that the Z translation (normal to the symmetry plane) be **FIXED**. The X and Y translations are both **FREE**. Also because of symmetry, we need to set the X and Y rotations as **FIXED**, and **FREE** the rotation around Z. Think carefully about these constraints (especially the rotations) and how they arise from symmetry. Accept the dialog.

The model is not yet fully constrained. We will also constrain the edges of the side and top inlet/outlet pipes. Select

> *Displacement*

again. Name the constraint *[sidepipe]* (still in ConstraintSet1). Set the **References** type to *Edges/Curves* and pick the outer (magenta) edge of the pipe coming out the side of the tank. Set all degrees of freedom to **FIXED** for this edge (no translation, no rotation). Repeat for the pipe leaving the top of the tank (call it **[toppipe]**). Zoom in to see the shape of the constraint icons along each edge.

Notice that the model is now overconstrained. That is, we have defined more constraints than are required to prevent rigid body motion. Can you list the redundant constraints? These are not "wrong" as long as they are consistent with our intentions for allowed deformation of the model. In this case, we are assuming that the two pipe connections to the environment are immovable and inflexible.

Since the display is getting a bit cluttered, use the *Simulation Display* options to turn off the load and constraint values. Observe carefully the information contained in the constraint icons along each edge.

Assigning a Pressure Load

Now, select

> *Pressure*

---

[2] If you select the *Intent* option for picking the edges, the edge-selection procedure is somewhat simplified. There will be 6 intent edges. You end up with the same edges highlighted in bold green, and the constraints are still the same.

Name the load *[presload]*, in **LoadSet1**. Check
the *Intent* option under **References** and click on
one of the interior surfaces. All of the interior
surfaces will highlight in green[3]. Enter a load
magnitude of **0.1** (note the default units are
MPa; our applied pressure is equal to 100 kPa,
about atmospheric pressure). This is gage
pressure since the pressure on the outside
surface is 0. Accept the dialog with a middle
click. The model should now appear as shown
in Figure 5 (load display arrows set to **Heads
Touching**).

You might as well save the model at this time.

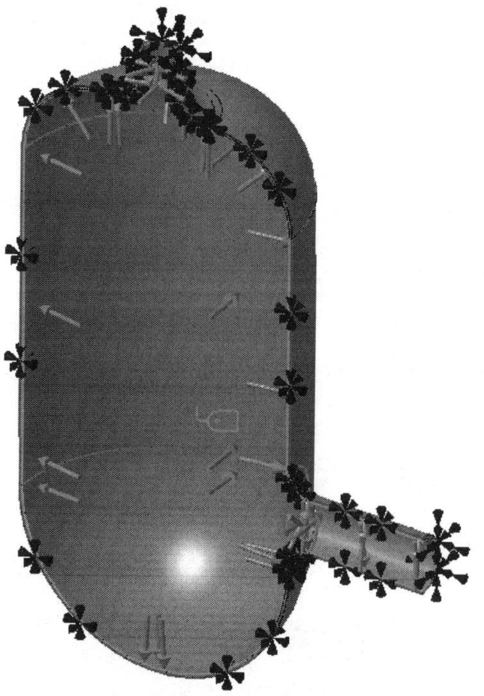

Defining and Running the Analysis

We can now define the analysis:

Figure 5 Completed model

Analyses and Studies > File > New Static

Enter a name *[shelltank]* and a description. Make sure constraint and load sets are
selected. Select a **QuickCheck** convergence. Review the **Run Settings**, and **Start** the
analysis. Always accept error detection for the first run of a new model. Open the
Summary window. About 50 shell elements are created[4]. The maximum Von Mises
stress is around 18 MPa and the maximum displacement is about 0.005 mm. Create a
result window showing an animated deformation to confirm that the constraints are doing
what you want. Assuming no errors, change the analysis to a **Multi-Pass Adaptive**
convergence (5% on **Local Displacement, Local Strain Energy & Global RMS Stress**,
max order 9) and rerun the analysis. You can use the elements from the existing study.
Open the **Summary** window. The run should converge on pass 7 or 8 with a maximum
Von Mises stress of about 17 MPa and a maximum displacement of 0.0091 mm. This is
one of the few times you may see the stress decrease from the QuickCheck to the MPA
analysis. Just above the reported measures, examine the values of the components of the
resultant load on the model. Can you explain/verify these values? How are these related
to the support reactions at the two pipes? You might like to create some measures for
those reactions to find out.

---

[3] You can, of course, pick these individually. There are 9 separate interior
surfaces, including the interior surfaces of the rounds. You may have to spin the model
to ensure that all surfaces are picked (it also helps if you are in shaded mode).

[4] Different builds of Creo may produce slightly different meshes, and
corresponding small changes in the results.

Viewing the Results

Create some result windows for Von Mises stress and deformation animation[5]. Show the element edges. Two views of the Von Mises stress are shown in Figure 6. Observe the location of the stress hot spots in relation to the deformation. This is easy to do if you show the stresses on the deformed shape. Note the scale on the deformation. Is the deformation consistent with your expectations? Locate the position for the maximum stress. Observe the variation in size and shape of the shell elements over the mesh. Create a p-level fringe plot and find out which elements have the highest edge order, and think about whether this corresponds (or not!) to the maximum stress level.

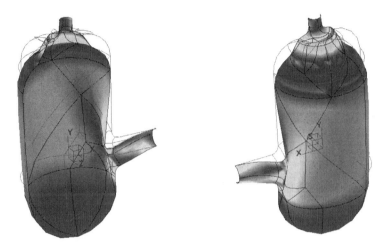

Figure 6 Von Mises stress fringe plot

The convergence behavior for this model is shown in Figure 7. This is a pretty well-behaved model.

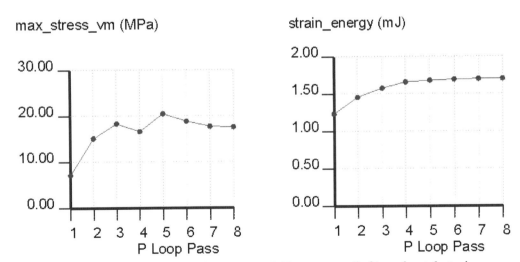

Figure 7 Convergence of maximum Von Mises stress (left) and total strain energy (right)

---

[5] You can use the previous MPA result window template for a solid model.

Exploring the Model

You should spend some time exploring this model. Here are a few things to try:

1. Delete (or *Suppress*) the existing constraint (symedges) on the cut edges of the shell then apply the *Symmetry Constraint* command (in the Constraints pull-down menu). The operation of the command is pretty self-explanatory. There are two types: mirror (default) and cyclic. We will look at cyclic symmetry a bit later. Select the same edges as before[6] and re-run the analysis. Do the results using this constraint agree?

2. In the *AutoGEM Settings* menu, there is an option for **Detailed Fillet Modeling**. Find out what this does in this shell model and what effect it has on the results.

3. Find out what happens to the convergence behavior and results if you change the AutoGEM limit for **Max Edge Turn** to 30°.

4. Back in Creo, change the radii of the rounds on the two pipes. How small can these be without significantly affecting the results? What happens (particularly to the convergence behavior) if you suppress the rounds altogether?

5. Modify the constraints on the side and top pipes. Remove all rotational constraints. For the side pipe, specify only translation in the X direction as fixed. For the top pipe, specify only translation in the Y direction as fixed. Coupled with the symmetry constraint (translation Z fixed), these are sufficient to remove all rigid body degrees of freedom. What effect does this have on the results?

When you have finished your explorations, return to Creo and save the model, then remove it from the session. We will move on to our next model.

This is a good time to take a break.

---

[6] If you have trouble with selected edges not being detected as in the same plane, you may have to change the system accuracy setting with *File > Prepare > Model Properties > Accuracy* and entering a new value of **0.0001** then regenerating the model. Or, without returning to the part window just use several constraints (autosym1, autosym2, ...) all within the same constraint set and containing different sets of edges.

Manual Shell Creation (Model #2)

Once again, the model (Figure 8) is created in Creo and we will use an idealization of the solid to create shell elements. In the previous model, we let Simulate automatically detect suitable surface pairs to form shell elements. This does not always work and/or you may want to be more selective. In this model, we will use a slightly different method to select the surface pairs that will be compressed to form the shell surfaces.

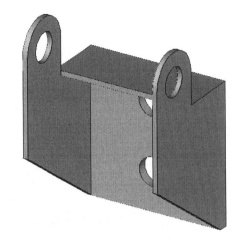

Figure 8 The mounting bracket model

Creating the Geometry

Create the model *[bracket]* according to the dimensions shown in Figure 9. Make sure your units are set to mm-N-s. The view of the part in Figure 8 is approximately in the default orientation. Note that there will be two different shell thicknesses (5mm and 10mm).

Assign the part material using *File > Prepare > Model Properties* and clicking on *change* on the top row. In the **Legacy** library, double click on **AL2014**, then *OK > Close*.

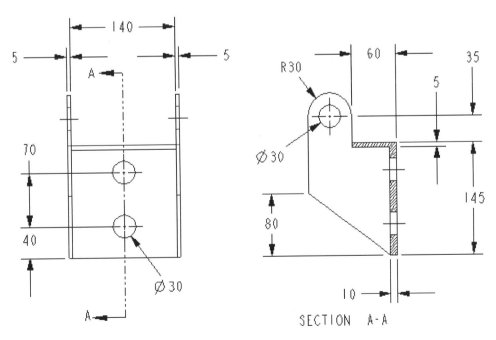

Figure 9 Dimensions (mm) of mounting bracket model

Defining Surface Pairs

When your model is ready, launch Simulate with

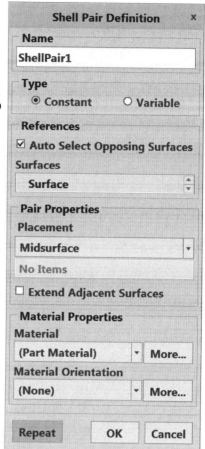

Applications > Simulate

The default model type is 3D. Open the **Simulation Display** dialog window, and in the ***Modeling Entities*** tab make sure that ***Shell Pairs*** is checked.

In the **Refine Model** ribbon, select

Shell Pair

We will create the shells manually using the dialog window shown in Figure 10[7]. As usual, we will accept all the defaults here. Note that this will create constant thickness shells, on the midsurface, using the current part material that we assigned earlier. We will let Simulate select the opposing surface for each one we pick ourselves (see the checkbox for **Auto Select Opposing Surfaces**).

Each shell pair will have a unique name - note the default at the top - starting with **ShellPair1**. With the **Surfaces** collector active, pick the outer vertical surface of the plate on the right side of the bracket, then middle click. The opposing surfaces of the pair will highlight in orange and magenta edges and the shell pair will appear

Figure 10 Dialog window for manually defining shells

under **Idealizations** in the model tree. The dialog window re-sets - note the new name **Shellpair2** in the collector at the top. Pick the outside vertical surface of the plate on the left side, then middle click. This creates the second shell pair. Pick the upper horizontal surface and middle click for the third pair, and then the front vertical surface of the rear plate and middle click for the final pair. The model tree should show four shell pairs under **Idealizations**. Close the shell definition dialog window with ***Cancel***.

In the AutoGEM group, select

Review Geometry > Apply

The midsurfaces will show in green. Click on a shell pair in the model tree and observe the display. Close the **Simulation Geometry** window.

---

[7] If you try to use ***Auto Detect*** exactly as we did for the tank, you will find that it won't work here. You must turn on the ***Use Geometry Analysis*** option and specify a thickness greater than 10mm.

Examining the Mesh

Open the *AutoGEM* pull-down menu. The option *Solid/Midsurface* is now checked. Change this to *Solid*, then select

AutoGEM > Create

AutoGEM will make about 270 solid tetrahedral elements. All of these span the thickness of the bracket plates. If there is any bending involved in these plates, we could anticipate that the element orders would have to be high to catch the stress gradients through the plate. So, examining this model using solid elements would likely be time consuming. Close the AutoGEM windows and do not save the mesh.

Now switch the *AutoGEM* pull-down menu to *Midsurface*. Open the Settings and check the box beside *Reentrant Corners*. Create the mesh. This time there are about 45 elements. Note that these are a combination of triangles and quadrilaterals (colored differently); we can force triangles only using an AutoGEM setting. Try this on your own later. The important point is that since shell elements explicitly allow bending in their formulation, we should expect the solution can be obtained with lower order elements. So we should save on two fronts in obtaining our solution: significantly fewer elements, and lower polynomial orders required for convergence.

Close all the AutoGEM windows. Do not save the mesh.

Completing the Model

Now apply constraints. As before, we will apply these to edges. Although this geometry is symmetric about the vertical midplane, the loading will not be (see Figure 12). Therefore, we can't use symmetry for this model. We will constrain the hole edges against translation only. In the **Home** ribbon select

Displacement

Create a new constraint set called *[fixededges]*. Name the constraint *[holes]*. Set the **References** type to *Edges/Curves* and pick on the front edges of the two holes on the back vertical plate of the bracket. Make sure you pick both halves of each circular curve. They will highlight in green. Leave all translation degrees of freedom FIXED and the rotations FREE. Accept the dialog with *OK*.

Now apply the loads. Use the ribbon button

Force/Moment

Create a new load set called *[holeloads]*.
Name the first load *[right]*. Select the
References type *Edges/Curves* and pick on
the outside edges of the hole on the right
vertical plate (both halves). They will
highlight in green. Select the *Advanced*
button and confirm that *Total Load* and
Uniform are the defaults. Set the X component
to *100*. Accept the dialog.

Follow the same procedure for the hole in the
other vertical plate. Name this load *[left]* (in
load set **holeloads**). Select the edge of the
hole on the left vertical plate. Select a *Total
Load*, *Uniform* distribution. In the **Force**
pull-down list, select *Dir Vector & Mag*. We
want a force 30° below horizontal in the YZ
plane, so enter the vector components **(0, -0.5,
0.866)** in the X, Y, and Z directions,
respectively. Enter a magnitude of **250**. See
Figure 11. Accept the dialog.

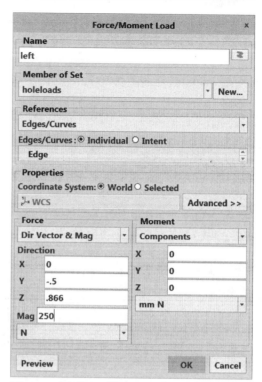

Figure 11 Specifying a force by
magnitude and direction

Note that we applied the loads and constraints
to edges, not surfaces (why?) and that we did not have to specify a shell property
(thickness) - this is obtained from the Creo solid model. Delete the empty load and
constraint sets in the model tree by selecting them and using the RMB pop-up. The model
is now complete and should look like Figure 12. You can change the appearance of the
model using the *Simulation Display* command in the Graphics toolbar.

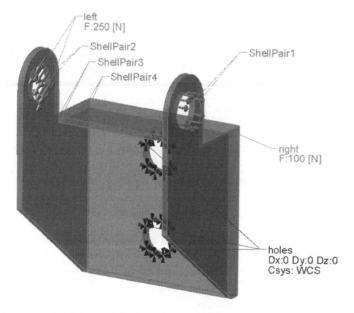

Figure 12 Mounting bracket model complete

Running the Model

Perform the usual analysis steps: set up a *New Static* design study and run a **QuickCheck** analysis. If you haven't deleted the empty sets, make sure you select the correct constraint and load sets. AutoGEM will create something like 45 shell elements. Open the **Summary** report. The maximum Von Mises stress is about 22 MPa. Note that several maximum stress measures have asterisks, indicating they were evaluated at or near a singularity. We can expect trouble ahead. You might like to create a result window showing a Von Mises fringe plot with element edges and an animated deformation. This will indicate whether the constraints have been implemented correctly and where the potential problems with singularities might lie.

Edit the analysis to run a **Multi-Pass Adaptive** analysis (5% convergence, maximum edge order 9). The multi-pass analysis does not converge on pass 9 - another indication that something is wrong. The maximum Von Mises stress has increased to about 60 MPa which seems a little high (more than 3 times the QuickCheck value). If you look at the data for Pass 9, you will see that a couple of elements have not converged.

Create some result windows to show the Von Mises stress and the deformation animation. The deformation is shown in Figure 13. Note the scale of the display. The maximum displacement is only about one-half millimeter. This looks fairly reasonable.

Now bring up the display of the Von Mises stress. It first appears as in Figure 14. Almost the entire model is shown in the lowest 2 or 3 fringe colors (below 20 MPa). There are two very small "hot spots" at the corners of the vertical plates. Show the location of the maximum Von Mises stress using

Model Max

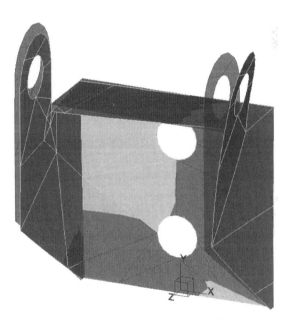

Figure 13 Deformation of the bracket

It occurs at the top corner on the right plate. To see the rest of the stress distribution in the part, we need to redefine the stress levels assigned to the colors in the legend.

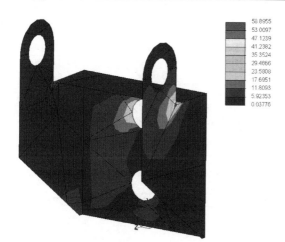

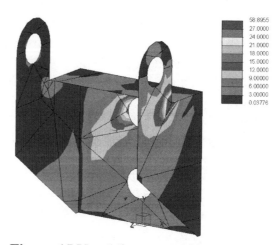

Figure 14 Von Mises stress fringe plot with default legend - not very useful here!

Figure 15 Von Mises stress fringe plot with edited legend - much better!

In the **Format** ribbon, select

> *Edit*

Change the values to read

Maximum	**27**
Minimum	**3**
Levels	**9**
Color Spectrum	**Creo Simulate Classic**

then select *OK*. Notice the combination of the minimum and maximum values and number of levels in the legend (3 x 9 = 27). This produces "nice" legend values that are easier to read. We have also put all the fringes in the lower half of the stress scale. The "hot spots" are now much more visible, as well as the stress distribution around the mounting holes. See Figure 15. Feel free to experiment with other settings for the legend levels and fringe color spectrum.

Create windows to show the convergence history of the maximum Von Mises stress and the total strain energy. These are shown in Figure 16. The Von Mises stress is increasing steadily with each pass with no sign of converging at all (although it looked like it might be converging on pass 3 and 4), while the strain energy does seem to be converging. This behavior coupled with the stress contours indicates that there is a singularity at the sharp inside (concave) corner. The p-code method in Simulate is not able to converge on this type of geometry, as it will continue to try to increase the polynomial order indefinitely in order to catch the (theoretically) infinite stress at the corner[8]. This means that any results (especially the stress) reported right at the corner

---

[8] For that matter, mesh refinement in h-codes could not produce a converged solution here either - the stress at the corner would continue to increase with each finer mesh.

(and in the immediate vicinity) must be taken with a large grain of salt - it cannot be trusted at all! This geometry is called a *re-entrant corner*. The presence of re-entrant corners and their effect on convergence analysis is a major complication for the FEA analyst. You can close the result windows.

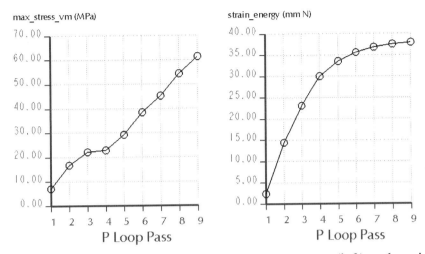

Figure 16 Convergence graphs of Von Mises stress (left) and strain energy (right)

Figure 17 below shows a closeup of the mesh created at the left re-entrant corner. To see this mesh, go to the AutoGEM menu and select *Create*. You can also select

Simulation Display > Mesh

and turn on the option to shrink elements. AutoGEM creates several small elements at this re-entrant corner on the left side. Inexplicably, AutoGEM does not create these small elements at the right side - it probably should since that is where the hotspot is.

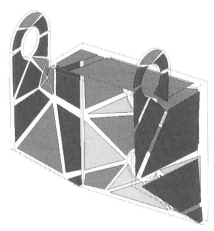

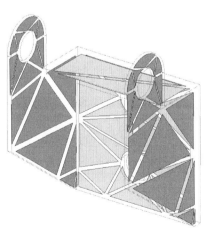

Figure 17 Default mesh showing elements at left re-entrant corner

Figure 18 Mesh created with **Re-entrant Corner** option turned *OFF* and "tri" elements only

In the AutoGEM settings dialog, set the following options:

Turn *ON*	**Move or Delete Existing Points**
Turn *OFF*	**Reentrant Corners**
Select	**Element Type: Shells(Tri)**

The first option causes all traces of any previous mesh to be eliminated when a mesh is deleted. Otherwise, node points are left in the model that will be incorporated in the new mesh. Turning off the second option tells AutoGEM to ignore potential re-entrant corners and just produce a simple mesh. The third option forces creation of triangular elements only. Return to the main AutoGEM menu and select *Create* again. This time there are only 56 elements. In Figure 18 notice the absence of small elements at the corner.

Rerun the MPA analysis to see what happens to the model results with this new mesh. What happens to the maximum stress? Convergence? If results have changed, explain why. If they haven't changed, explain that too!

Using Excluded Elements

It is not clear why AutoGEM recognizes the re-entrant corner on the left side but not the right side. The methods described above (along with another procedure in one of the exercises) are possible ways to deal with the presence of a singularity in the model. Earlier in these lessons, we saw that excluding elements from the convergence analysis is also a possibility. You will recall that this involved creating an AutoGEM control for mesh creation and then specifying in the design study that results on these special elements should not be used in the convergence analysis - they should be excluded. Let's see how that works.

In the **Refine Model** ribbon, AutoGEM *Control* pull-down list select

Isolate for Exclusion

Select the button ***Preselect Singularities***. Another window opens. Turn on the option **Re-entrant Corners Less than 120 deg**. Now press the ***Select*** button. Nine edges are highlighted (they are all less than 120 degrees). We do not want all of them, so holding down the CTRL key, deselect the two long vertical edges and the horizontal edge at the back. This will leave six edges in the collector[9] and highlighted on the model. In the **Isolation for Shells and Boundary Mesh** drop-down list, select **Automatic**. Make sure the **Exclude** box at the bottom of the window is checked. Middle click to accept the control definition. The model now shows the excluded control icons along the six edges. See Figure 19. Note the appearance of this control in the model tree.

---

[9] Expand the collector to see the six edges. If a surface is also shown, delete it.

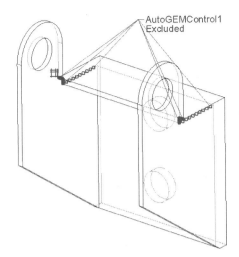

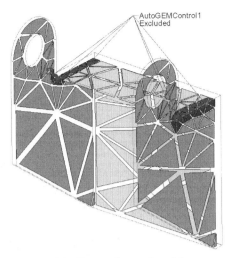

Figure 19 Selected edges for AutoGEM control

Figure 20 Created mesh with Isolating Elements shown in red

In the AutoGEM window, select **Create**. Something like 180 elements will be created[10]. In the **Simulation Display** controls, you can shrink the mesh elements to make them easier to distinguish. Back in the AutoGEM window menu, select

Info > Isolating Elements

Any elements in contact with the edges selected previously will be shown in red (Figure 20). Note that there are a few more elements near the re-entrant corner on the right vertical shell than for the default mesh (compare with Figure 17).

Go back and **Edit** the MPA design study analysis definition. Select the **Excluded Elements** tab and turn on the check box beside **Exclude Elements**. The default action is to ignore stresses. Note that we can also limit the polynomial order in these elements - leave that at default (6) for now as well.

One of the result windows we will look at for this model will be a graph of Von Mises stress values along a couple of the model edges. We would like to get some higher definition for this graph data. Select the **Output** tab, and change the **Plotting Grid** to **8**. This increases the number of points along each element edge where the results are computed for display.

Leave the convergence settings as they are (5%, max order 9) and **Run** the study. The solution will converge on pass 8 (it gets just under 5% on the **Global RMS Stress Index**). In the reported measures you will see that the maximum Von Mises stress is around 49 MPa. As we will see in a minute, this is **NOT** the model maximum, since that occurs on one of the excluded elements. The measures reported in the **Run Status** report are for non-excluded elements. None of the stresses have the asterisk and there is no warning about measures taken at or near singularities. Most of the measures have converged very well as indicated by the percentage values in the right column.

---

[10] **Re-entrant Corners** = **OFF** and **Element Type** = **Tri** as previous.

Create a Von Mises stress fringe plot.
You will have to edit the legend, since it
is stretched to the maximum value
reported of around 250 MPa. Set the
values as we did before (***Format > Edit***,
min 3, max 27, levels 9). See Figure 21.
With this legend, most of the fringes are
in the lowest 20% of the stress values.
There are two "hot spots" at the re-
entrant corners, but these have not upset
the convergence of the model. Use
Dynamic Query to find the location of
the reported maximum value of 47 MPa -
you should find it at a point closest to the
re-entrant corner on the edge of one of
the non-excluded elements. The ***Model
Max*** will show the stress at the corner.

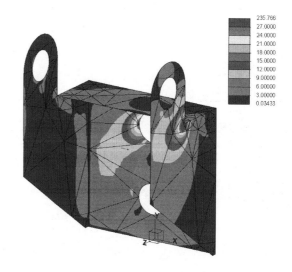

Figure 21 Von Mises stress in model with
excluded elements

Check out the stresses reported on the left re-entrant corner. There is a singularity here as
well. Due to the current loading, it is just not as pronounced. You might also consider the
stresses reported around the mounting holes. These values are probably fairly accurate,
although we note that values reported on the locations where we apply constraints must
be considered carefully.

See the convergence graphs shown in Figure 22. Not bad! We probably could have
stopped on pass 5 or 6 (using 10% convergence perhaps).

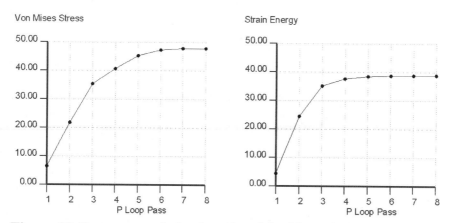

Figure 22 Convergence behavior of model with excluded elements

Although we have now gotten around the problem of the singularity upsetting our
convergence analysis, we are still faced with the question of what the value of stress at
the corner actually is. Due to the singularity, the reported numerical value is suspect.
Here is a useful way to obtain a better estimate.

Create a new result window. Set the *Type(Graph)*, *Quantity(Stress, Von Mises)*. In the Graph Location panel pick the button below **Curve**. A new window opens showing the model. Select the two edges on the right face that lead into the re-entrant corner. See Figure 23. When these are accepted with a middle click, you can set which end of the composite curve you want at the graph start (left end of graph), marked by a very small "+" sign. It is not critical here which one you use. Accept the dialog and show the window. A graph is created (Figure 24) that plots values of the Von Mises stress along the edges leading into the singularity. The smoothness of this graph is due to the density of the mesh and the value we set for the plotting grid earlier.

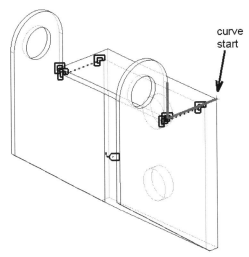

Figure 23 Selecting edges for a graph of stress values

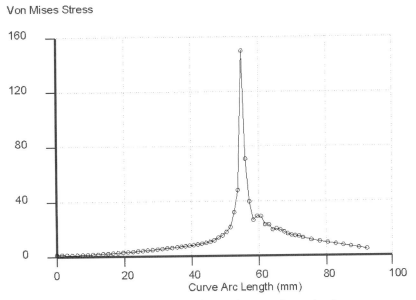

Figure 24 Von Mises stress values along selected edges

The circular symbols indicate the locations (8 per element edge) where the stress value is computed. It is clear that only the element directly at the corner is affected, and only over a fairly short distance. If you tried to refine the mesh even further, you would find that the sharp spike becomes taller and narrower - it is clearly a numerical artifact and not a real physical effect. It is probably safe to use curves like this to estimate what is really happening at the corner. In the physical model, there are two real effects that must be considered. First, it is physically impossible to have perfectly sharp corners - all corners have some size of fillet or round. Second, if a fillet radius is small enough, there will be a stress concentration that could potentially lead to local yielding. This limits the stress and causes adjacent material to carry more of the load. This is a non-linear type of behavior that can be treated by Simulate by setting the material type to "elastoplastic".

You might like to experiment a bit with the AutoGEM control for setting up the excluded elements. For example, set the maximum element size to 5mm on the edges coming in to these corners. This creates over 300 elements. The reported maximum Von Mises stress in the study report will be around 50 MPa (since the location is closer to the corner) although the Von Mises fringe plot shows a maximum of 250 MPa. The hot spot for the high stress area at the corner is smaller, and the stress spike along the curves (as in Figure 24) is even sharper (taller and narrower).

Close this model and take a break!

Mixed Solids and Shells (Model #3)

Quite often, a solid part will contain regions of thin-walled material. The part could perhaps be modeled completely using solid elements. However, it will usually be more efficient to model any thin-walled features using shell elements, leaving the rest of the model as a solid. This is illustrated by the example in this section. We will also look at the *Bearing* load, and some new forms of boundary constraint.

Figure 25 Bell crank model

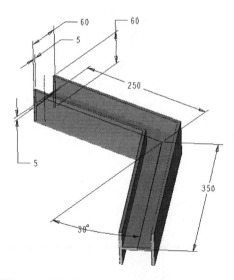

Figure 26 Base feature - Sweep

The part we will model, a simple (large!) bell crank, is shown in Figure 25. The base feature for this part is a simple sweep, shown in Figure 26. The dimensions for the part (in millimeters) are shown in Figure 27. Start this new part called *[crank]* using the mm-N-s part template. Notice the orientation of the default coordinate system in Figure 25. The sweep can be created using a sketched trajectory on the TOP datum plane. The central of the three bosses at the corner of the sweep is located at the origin. After the sweep, create the three bosses and then coaxial holes (all *Through All*).

Assign the material STEEL to the part using *File > Prepare > Model Properties*.

When the model is completed, transfer into Simulate with

Applications > Simulate

Remember that the default model type is 3D, so we don't need to do anything about the model type.

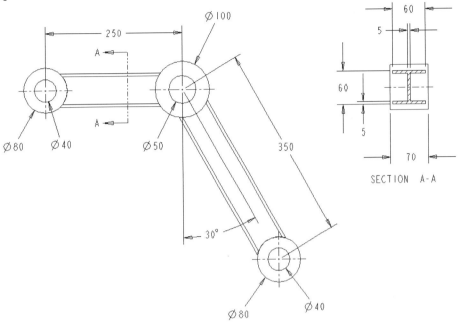

Figure 27 Dimensions of the bell crank model

Creating the Shells

The first thing to do is identify where we want shell elements. In the **Simulation Display** panel, open the **Modeling Entities** tab and make sure **Shell Pairs** is turned **ON**.

We can either use automatic shell creation (as in Model #1) or semi-automatic creation (as in Model #2). Since we can use a variation of automatic creation, we will go that route (come back later and try the other method). In the **Refine Model** ribbon, *Shell Pair* pull-down list select:

Detect Shell Pairs

Make sure the option *Use Geometry Analysis* is checked and enter a **Characteristic Thickness** of *10* mm. This means that all sets of parallel surfaces less than 10mm apart will be chosen to create shell pairs. In this model this will capture all the shells we want. Click on *Start*. The message window indicates that 5 shell pairs have been created - these are listed in the model tree under **Idealizations**. These appear as green and cyan surfaces. If you *Repaint* the screen, the paired surfaces show in orange and magenta edge highlights. Selecting a shell pair in the model tree will display the surfaces in green and cyan.

Go to the pull-down AutoGEM menu to make sure that the *Solid/Midsurface* option has been checked - we need both for this model.

In the AutoGEM group, select *Review Geometry*. Turn *ON* the checkbox beside *Shell Surfaces*, and *OFF* the checkboxes beside *Solid Surfaces* and *Original Geometry* then click on *Apply*. The display will show only the defined shell surfaces in green (Figure 28). If you turn *OFF* the *Shells* and turn *ON* the checkbox for *Solid Surfaces* and then *Apply*, just the three bosses will be displayed (Figure 29). Close the **Simulation Geometry** window.

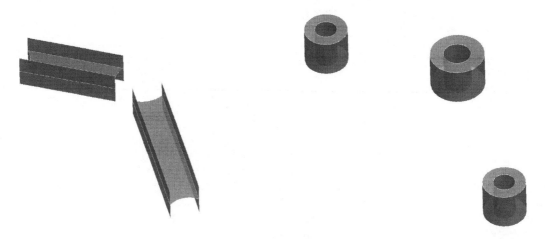

Figure 28 Model with Shells only   **Figure 29** Model showing Solids only

We can now proceed to define the rest of the model.

Defining the Constraints

For the constraints, we want to restrict motion of the central boss and the one at the end of the long crank arm. We will constrain each hole surface against radial and axial motion but allow rotation around the hole axes. To do this, it is most convenient to use a new type of displacement constraint (an alternative method is described below).

In the **Home** ribbon, select the *Pin* constraint button (this looks like this). This opens the dialog window shown in Figure 30. Name the constraint *[pins]* and pick (with CTRL) on the hole surfaces on the middle and the right end bosses. You might note that preselection will only highlight cylindrical surfaces here - those are the only type of surfaces allowed for a Pin constraint. In the dialog window, set the angular constraint to *Free* and the axial constraint to *Fixed*. This allows the surface to rotate as if it was mounted on a shaft but not translate along the shaft. There is an implicit constraint against radial displacement of the surface. This also implicitly restricts the right arm of the crank from stretching (the two pinned surfaces cannot get farther apart). When you accept the dialog, the constraints will appear as in Figure 31.

Figure 30 Creating a
Pin Constraint

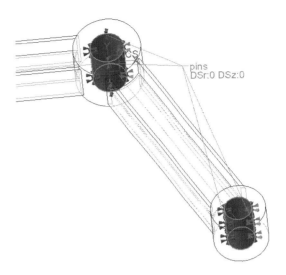

Figure 31 Pin Constraints defined on holes

An alternative to the **Pin Constraint**, that would do the same job but is not restricted to cylindrical surfaces, is as follows. This involves the creation of a cylindrical coordinate system centered on each hole, with the Z-axis of the system lined up with the hole axis. You would then use the usual displacement constraint tool and specify that the constraint was defined in the cylindrical system. When you switch the definition from a Cartesian system to a cylindrical system, the constraint directions will automatically switch from **X-Y-Z** to **R-Theta-Z** directions. To duplicate the pin constraint, you would then **Fix** the **R** and **Z** displacement, but **Free** the **Theta** displacement. You would need to set up constraints for the two holes separately, using each hole's cylindrical coordinate system as reference. This is a bit more work than necessary for simple pins but is more generic and might be useful in other situations.

Defining a Bearing Load

We will apply a bearing load on the hole at the end of the shorter crank arm. This is another special-purpose tool similar to the pin constraint used above. A bearing load can be applied only on cylindrical surfaces (interior or exterior) and has a resultant force in a specified direction. The actual force is applied normal to the bearing surface in a non-uniform distribution, more or less to model what would happen if a solid shaft were placed in a hole and a lateral force applied on the shaft in the direction specified.

In the **Home** ribbon, select

> ***Bearing***

This brings up the dialog window shown in Figure 32. Name the load ***[bearing]*** (in LoadSet1). Click on the hole surface. Note the default coordinate system (WCS). We can specify the direction in three ways. In the **Force** pull-down list, select **Dir Vector & Mag**. Enter the data shown in Figure 32. Finally, enter a magnitude of **1000**.

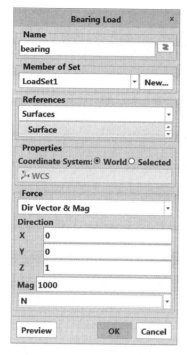

Figure 32 Defining a Bearing Load

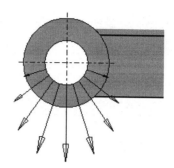

Figure 33 Preview of the Bearing Load

Preview the load to see how it is being applied. This is best seen in Top view and gives you some idea of how the load is distributed. Note that it applies a compressive load normal to the surface of half the hole and no load on the opposite surface. This distribution will reorient with the direction vector. The lateral forces on the bearing surface exactly cancel out, while the resultant force is as specified. Accept the dialog. The bearing load icon will be somewhat different in the normal display. The model is now complete.

Running the Analysis

Create a *New Static* analysis called *[crank]*. As usual, start with a **Quick Check**. Verify your *Run Settings* before selecting *Start*. Open the **Summary** window. AutoGEM creates something like 38 shell elements and 200+ solid elements with reentrant corners turned off and quad/tri elements. Note that AutoGEM creates entities called **Links** - these are necessary to connect shell elements (which have rotational degrees of freedom) to solid elements (which do not) in order to transfer load correctly between these two types of elements. The maximum indicated Von Mises stress is about 16 MPa. You should have a look at the results of the Quick Check (especially the deformation) to make sure the model is doing what you expect it to. Notice the rotation of the bosses. If all is well, *Edit* the analysis to set up a **Multi-Pass Adaptive** analysis with the default 10% convergence and maximum edge order at 6. Run the MPA analysis.

Reviewing the Results

The MPA analysis does not converge. The maximum Von Mises stress has increased to about 20 MPa. In the **Summary** window, the convergence for many stress components is larger than 20%. This indicates either that we should modify the mesh parameters to have AutoGEM create more elements, and/or we have something like a singularity condition in which case we should consider excluding elements.

Create the usual result windows to show a fringe plot of the Von Mises stress, a deformation animation, and the convergence of the Von Mises stress and strain energy. Also, create a fringe plot showing the P-levels of the mesh.

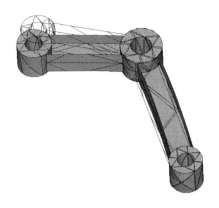

The deformed shape is shown in Figure 34. Observe that the constraints are doing what we want - the bosses can rotate around their axes, but cannot move axially or radially. You might have to zoom in on the end boss to see the rotation.

Figure 34 Deformation of bell crank

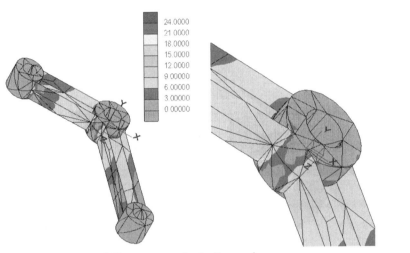

Figure 35 Von Mises stress in bell crank

The Von Mises stress is shown in Figure 35. Find out where the maximum stress is located. This may be the location of our convergence problem.

The convergence graphs are shown in Figure 36. There is clearly a problem with this model in regard to the maximum Von Mises stress, although it is likely that the rest of the model is satisfactory. The strain energy has almost converged (more or less) on the 3rd pass, but the stress keeps rising. This is a sign of a singularity condition and it is difficult, therefore, to say anything conclusive about these results.

Finally, have a look at the P-levels in the mesh. Where are the highest order elements? Do these correspond to the stress hot spots? If so then we have further evidence of a singularity-like condition.[11]

---

[11] The author pondered a long time about including this exercise in the lesson since it may leave the reader with a rather low opinion of this type of model. It was decided to include it specifically because it displays some of the diabolical behavior that

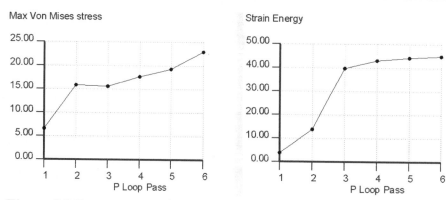

Figure 36 Convergence of Von Mises stress (left) and strain energy (right) for the bell crank model

Exploring the *Thin Solid* option

As an alternative to the shell element, a special way of treating solids with thin cross sections is available with the ***Thin Solid*** mesh control option. Go to the model tree, and use the RMB pop-up menu on **Shell Pairs**. Then select

Convert to Thin Solid AutoGEM Control

This removes the shell pair definitions and replaces them with an AutoGEM control[12]. Go to the **Refine Model** ribbon, and use ***AutoGEM > Create*** to produce a new mesh. This consists of over 1000 solid elements and is a mix of tet, wedge, and brick elements as shown in Figure 37. This is one of the few times you will see the latter element types. Observe the shape of the elements formed in the thin wall of the section. Despite the ruggedness of p-elements, we might expect the required edge orders to be high. With this many elements, then, we would expect a significantly longer run time than we saw before.

Figure 37 Mesh created with **Thin Solid** control

must be dealt with by the FEA analyst! FEA is not always as straightforward as many people think, and many models will give you problems with convergence. Notwithstanding the performance of this model, mixed solid/shell models are very useful and often necessary.

[12] The ***Thin Solid*** control is also available in the AutoGEM control pull-down list in the **Refine Model** ribbon.

Go back and re-run the last MPA study (10% convergence, max order 9). This claims to converge on pass 6 (with max edge order 8), with warnings about stress values being computed near singularities. Have a look at the usual result windows. The MaxVM convergence result window may be of particular interest! Has the location of the max Von Mises stress also changed?

Summary

This lesson has introduced you to the main tools for using shell pairs in general 3D models. For certain models, these surface pairs can be selected automatically. Otherwise, you can select paired surfaces manually. In either case, the shell thickness is read from the actual Creo model. Shells can also be defined on single model surfaces, in which case the shell thickness is specified rather than read from the model, similar to what was done for axisymmetric shells (or plane stress models). It is also possible to create shells with step-wise or continuous variation in thickness. Consult the on-line help for further information on these advanced topics.

One of the main problems you will face with shell models is re-entrant corners. If present, these may dominate the convergence behavior. Even worse, they leave you guessing at the correct stress levels in precisely the areas where they are highest. It might be possible to modify the mesh somewhat near these singularities using the mesh control methods discussed in a previous lesson (datum curves, edge controls, datum points, AutoGEM limits, etc.). Another possible way around the singularity problem is to monitor convergence at points removed from the singularity location, perhaps by seeding the geometry with datum points. Measured quantities can be set up at these points and monitored for convergence, rather than looking at the model maximum. Perhaps the easiest way to deal with singularities is using excluded elements. It may still be a challenge to evaluate stresses at critical locations.

We also saw how a bearing load can be created and some variations on specifying constraints that allow rotation around an axis (the *Pin* constraint).

Despite the somewhat dubious performance of our third model, shell elements can be very beneficial in models with thin-walled features. They can drastically reduce the number of elements in the model and the computation time to get a solution. As always, we must be very careful about interpreting these results.

In the next lesson, we will look at another of the idealizations covered in this book. This idealization is used in the treatment of beams and frames.

Questions for Review

1. When will **Detect Shell Pairs** locate surface pairs of thin-walled features? (For what Creo features and geometry will it *not* work?)
2. What does the option **Use Geometry Analysis** do?
3. The **Shell** command in Creo Parametric can be used to create shells with different thicknesses on different surfaces. How does this affect **Detect Shell Pairs**?
4. Suppose you have a model with a number of thin-walled features (say 1.0mm thick) and also have several parallel surfaces slightly thicker (say 10 mm thick) that you want to treat as a solid. How does **Detect Shell Pairs** behave in this situation?
5. Can you delete a single shell pair from a set found using **Detect Shell Pairs**? What happens if you still try to execute the model?
6. What happens if, when selecting surface pairs manually, you accidentally pick two surfaces that intersect?
7. Can you manually pick concentric cylindrical surfaces as members of a shell pair?
8. How do you specify the material for a model composed entirely of shell elements?
9. What type and shape of elements does AutoGEM create for general 3D shells?
10. Is it possible to create a part model with two shells that pass through each other?
11. What are the restrictions on applying loads and constraints to shell models?
12. How do you specify the thickness of a shell model? Does this have to be the same everywhere in the model?
13. An end view of a small portion of a solid part is shown in the figure. Each of the three thin "spokes" extends for some distance into the page. We want to model each spoke as a shell. Do you anticipate problems? If so, what?
14. Suppose you are making a model of a thin-walled part, which must have draft on all vertical surfaces. How will this complicate your model?
15. What are the symptoms of the presence of a singularity in the FEA model? How will this complicate the analysis of the results of your model?
16. Normally, if a model will not converge with the maximum edge order (9), the solution is to modify the mesh to produce more, smaller elements (mesh refinement). Will this strategy work in the presence of a singularity? Why?
17. If the maximum Von Mises stress result is not suitable for monitoring convergence in a model with a singularity, what other measures could you use to monitor convergence? (See also exercise #4.)
18. What is the display icon that represents a **Symmetry Constraint**?

Exercises

1.   A U-shaped steel beam is cantilevered out from a wall and can carry a wide range of loads in the X, Y, and Z directions shown. The beam is a C-channel with a thick plate welded on the end. It is desired to set the model up so that, using superposition, the solution can be found for any loading scenario using the three components.

A solid model (*ubeam.prt*) can be downloaded from the SDC web site. It will be hopeless to try to treat this model as a solid. Use shells instead and defeature the model as required. Choose from the methods described in this lesson to create the shell idealization from the solid model.

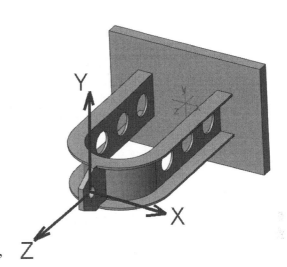

Hint: There may be singular locations on this model and you will have to be creative in finding ways to monitor convergence. Due to singularities, the reported values of the maximum stresses are probably suspect, so you will have to again be creative in using the computed stress values in the model to estimate the actual maximum stress.

2.   Make a solid model of the centrifuge used in Chapter 6, as shown in the figure (a one-quadrant model due to symmetry about three planes). (HINT: in Creo use a *Thin, Revolved Solid*). Bring this into Simulate and model it using 3D shells. Can you use *Detect Shell Pairs*? The material is **AL2014**. Apply a centrifugal load of 8000 RPM. Set up an optimization to find the

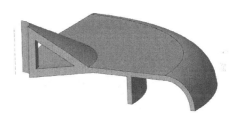

minimum mass without exceeding a Von Mises stress of 50.0 MPa on the extreme outer rim. (Hint: Beware of values of the model maximum Von Mises stress since that may be influenced by singularities.) The design variables are the thickness of the shells (*thick5* and *thick8* in the previous model) and the radial location of the inner vertical rib support. How much was the mass reduction from the initial design? Show this in a graph (mass vs optimization iteration). The ranges for the design variables (in mm) are as follows:

	Minimum	Maximum	Initial
thick5	2	6	5
thick8	4	10	8
rib location	100	160	130

3.   Continue the optimization of the previous question adding the following factors: The exterior of the centrifuge is an evacuated chamber (to reduce air friction) while the inside of the centrifuge is pressurized to 30 kPa (absolute) as shown in the figure below. In addition, we cannot allow the maximum radial or vertical deflection anywhere to exceed 0.2 mm. Find the thickness of the shells and the location of the vertical support that will result in the minimum inertia, Iyy, around

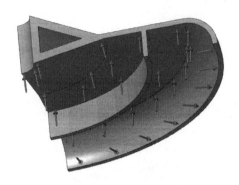

the axis of rotation. (Minimum inertia means that, for a given torque, the centrifuge will accelerate as fast as possible.) For the optimized design, which constraints are active?

4.   Consider the bracket model again (model #2). We saw that the model's maximum Von Mises stress was an unsuitable measure for monitoring convergence when the model contains a re-entrant corner or other singularity. Create the datum curve (radius = 10mm) and point PNT0 shown on the part below. Use these in the appropriate AutoGEM controls (***Hard Curve*** and ***Hard Point***). Define a measure to monitor the Von Mises stress at this point during an MPA analysis and use it to determine

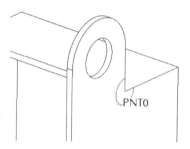

convergence. Compare these results to those obtained during the lesson and comment on the validity of this procedure.

5.   Create a simple hollow box using a protrusion and a cut as shown below. Note that each side (including the bottom if you make the interior cut 75 deep) of this box has a different thickness. Examine the behavior of the settings for automatic detection of shell elements obtained with ***Detect Shell Pairs***, turning on the option ***Use Geometry Analysis***, and using different thickness values.

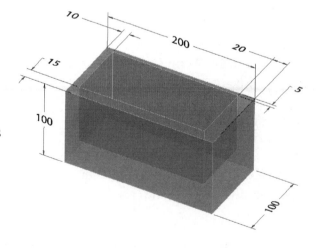

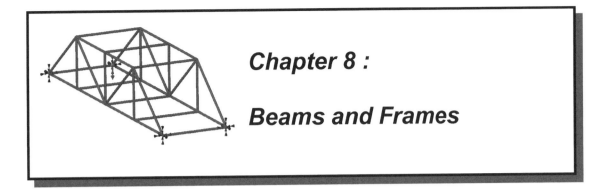

Chapter 8 :

Beams and Frames

Overview of this Lesson

Beams are fundamental structural elements. In Creo Simulate they are treated as idealizations. Beams are one-dimensional elements that can be either straight or curved. Due to the underlying mathematical theory and formulation of the element equations, the element length must be much greater than the cross section dimensions (a rule of thumb is 10 times) for good accuracy. The cross section must be constant along the beam element. A model might be composed entirely of beam elements, or they can be used in conjunction with other model entities (solids and/or shells).

This lesson introduces the main concepts required to model isolated continuous beams and beam elements as components of frames[1]. The subject of beam coordinate systems is introduced with sufficient depth to handle problems with simple symmetric beam cross sections. Beam orientation in 3D requires a solid understanding of these coordinate systems.

Four example problems are used to illustrate the Simulate commands. The first is a simple cantilever beam (a diving board) with a tip load. New results windows are created to show the shear force and bending moment diagrams for the beam. The second example is a more complicated continuous indeterminate beam. This introduces distributed loads and the use of beam releases. The final example is in two parts: a simple 2D frame and a 3D frame. These illustrate more ideas in beam orientation and gravity load. In the final model, the loading caused by a specified displacement of a constraint is used to model the settling foundation beneath one corner of the 3D frame.

Beam Coordinate Systems

One of the potentially confusing issues in using beam elements is their orientation with respect to the World Coordinate System (WCS). This is particularly true for curved beams, and beams whose cross sections are asymmetrical about their centroid in at least

---

[1] Interestingly, the most simple and basic of structural configurations to solve analytically (a truss) is the most difficult to set up with Simulate due to the special conditions that must be applied to the beam elements in the model.

one lateral direction (such as channels and angles) and/or are offset from the underlying geometric curves. For the most general case, this orientation is described/defined using up to three coordinate systems. For this lesson, we only need to worry about two of these - the BACS and the BSCS[2].

The Beam Action Coordinate System (BACS)

Beams are associated either with geometry curves or connecting two (or more) points defined in the WCS. The curves and/or points can be created either in Creo Parametric (as datums) or in Simulate (as simulation features). We will deal only with straight beams in this lesson. For these, the underlying curve or element end points will define the X-axis of the beam's BACS (see Figure 1). The beam's local Y- and Z-axes are perpendicular to the beam. The orientation (relative to the WCS), i.e. rotation of the beam around its local X-axis, is defined by specifying the direction that the BACS Y-axis is pointing. This direction can be specified in a number of ways: an axis direction, an edge, a point, or giving vector components in the WCS. The BACS Z-axis direction is obtained by the right hand rule. Some simple examples showing the specification of the BACS Y-Axis using vector components are shown in Figure 2. Notice the orientation of the WCS in the center of the figure. Although the figure shows Y_{BACS} parallel to one of the WCS coordinate directions, that is not required. The Y_{BACS} direction vector can point in any direction perpendicular to X_{BACS}. All the properties of each beam element are set up in the **Beam Definition** window. This includes the geometric references, material, orientation of the Y-axis, section shape, and so on. Several beams with the same properties can be created simultaneously.

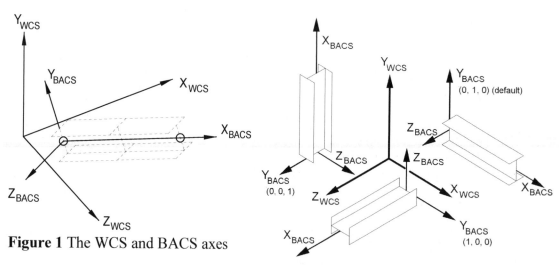

Figure 1 The WCS and BACS axes

Figure 2 Illustrating BACS Y-Axis Orientation (vector components given in WCS)

---

[2] The third system is the BCPCS (Beam Centroidal Principal Coordinate System). See the on-line help for further information on this system.

The Beam Shape Coordinate System (BSCS)

The beam cross sectional shape and position are defined relative to the BSCS. The standard cross sections built in to Simulate are shown in Figure 3. You can also create your own section shapes using Sketcher. The shape is defined in the BSCS YZ plane. For most standard shapes, the origin of the BSCS coincides with the centroid of the section. The X-axis of the BSCS (coming out of the page in Figure 3) is always parallel to the X-axis of the BACS, that is, along the beam. The BSCS origin (or its shear center) is defined by offsets DY and DZ measured from the origin of the BACS. See Figure 4. The orientation of the BSCS is determined by the angle theta specified in the Beam Orientation property window.

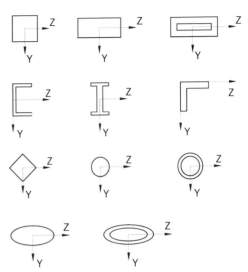

Figure 3 Standard Beam Section Shapes defined in the BSCS axes

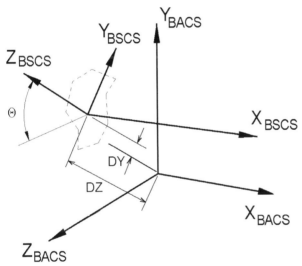

Figure 4 Definition of the BSCS axes relative to BACS. The frames coincide when theta, DY, and DZ are all 0 (this is the default)

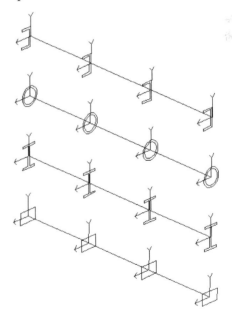

Figure 5 Beam elements with section icons (**Wireframe** display)

IMPORTANT: The BSCS is parallel to the BACS if theta is zero. In addition, if the offsets DY and DZ are zero, then the BSCS coincides with the BACS. This default will be the case in all the examples used in this lesson. **Note that DY and DZ are measured relative to the BSCS directions.**

When a beam cross section and orientation are specified, the combined properties will appear as an icon in true scale at four locations along the beam element. This serves as a visual cue to the shape, size, and orientation of the beam. Some examples are shown in Figure 5. These icons also indicate the directions of the BSCS Y- and Z-axes. The Y-axis is an open V (the Y axis is upward in Figure 5), while the tip of the Z-axis is an open arrowhead.

Now, this all seems pretty complicated, and indeed it is. Fortunately, in most cases, the Simulate defaults are exactly what is required and you can do a lot of modeling knowing only the bare essentials.

Example #1 - Basic Concepts

Our first example is a simple beam similar to a diving board. This will introduce some of the concepts involved in using beam elements.

The Model

The model is shown in Figure 6. The cross section is a hollow rectangle, 24 inches wide and 2 inches high, with a wall thickness of 0.125 inch. The beam is cantilevered out from the wall, and rests on a simple support 10 feet from the wall, with an overhang of 6 feet. The material is aluminum and a downward vertical load of 200 lb is applied at the tip. This is a static load, and we will calculate the shear force and bending moment and static deflection, as if someone was just standing at the tip, not bouncing up and down. Note that this problem is statically indeterminate - that is, the simple methods of statics cannot be applied because of an extra unknown in the reactions. Nonetheless, analytical methods[3] could be used to solve for such things as the bending moment along the beam and the deflection at the tip and, eventually, for the stresses in the beam.

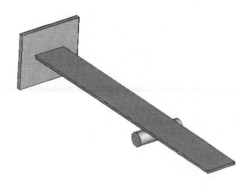

Figure 6 The diving board model - a simple indeterminate beam

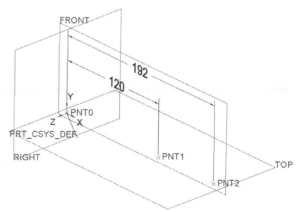

Figure 7 Datum points created in Creo

---

[3] See the *singularity method* discussed in most mechanics of materials textbooks.

Geometry

Create a new part in Creo called *[divingboard]*. You can use the default part template, but change your units (use *File > Prepare > Model Properties*) to in-pound-sec (IPS). The beam will be created along the X-axis of the default coordinate system. Since we are going to use an idealization of the beam, all you need to do here is create three datum points (perhaps using the *Offset Coordinate System* command on the *Point* dropdown). Create the three points at the locations (X, Y, Z) = (*0, 0, 0*), (*120, 0, 0*), and (*192, 0, 0*). See Figure 7.

With the datum points created, transfer into Simulate with

Applications > Simulate

Keep the model type default (3D).

Beam Elements

We will create two beam elements directly from the points. In the **Refine Model** ribbon, select:

Beam

The **Beam Definition** window appears. It will look like Figure 8 when we are finished. Accept the default name **Beam1**. The **Reference** type we want is **Point-Point**. Observe the reference collectors as you pick the point at the origin (**PNT0**) and the point that will be under the support (**PNT1**). Beside the **Material** pull-down list (currently empty), select *More*. Select the material **AL2014** then select *OK*. The default Y-direction is (0,1,0), that is, parallel to the WCS Y-axis - exactly what we need. Under the **Start** tab, beside the **Beam Section** pull down list, select *More*. This brings up the **Beam Sections** window where you can access a user-created library of beam sections. It is likely currently empty so select *New*, which opens the **Beam Section Definition** window shown in Figure 9. Name the section *[hrectangle]* and in the **Type** pull-down list select **Hollow Rect**. Enter the dimensions beside the sketch of the cross section as shown in Figure 9. The **Review** button in this window will calculate section properties for you and display

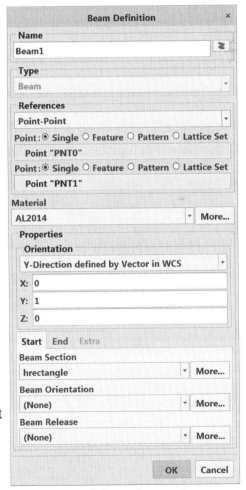

Figure 8 Defining a beam element

them in the Browser. Accept the dialog, then *OK* to get back to the **Beam Definition** window. Notice that **Beam Orientation** is set to (**none**). This refers to the BSCS offset and angle settings (DX, DY, and Theta) shown in Figure 4. By default, these are all zero and the BSCS aligns with the BACS. In the **Beam Definition** window, select *OK*. The first beam element appears as a shaded solid in green in true scale. Clicking on the background deselects the element and therefore changes from green to the color set by the button at the top-right corner of the **Beam Definition** window in Figure 9. The default is blue. This shaded display can be a bit misleading - it is not affected by the standard display settings (shaded, wireframe, etc.). If you go to the **Simulation Display** window (via the Graphics toolbar) you can select from the three display settings for beams (*Wireframe, Shaded, Transparent*).

Now create the second element. In the toolbar, select *Beam*. The definition window opens again. The default name is **Beam2**. All you have to do here is set the reference type to **Point-Point**, then select the new points (**PNT1** and **PNT2**). The material AL2014 can be selected from the pull-down list. Middle click to accept the completed dialog. Both beam elements are now defined; see Figure 10. Notice the section Y and Z directions on the section icons.

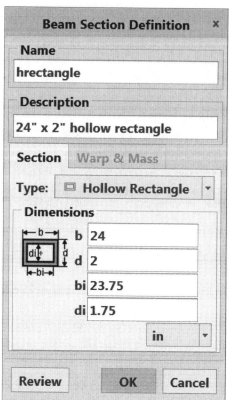

Figure 9 Specifying the beam section properties

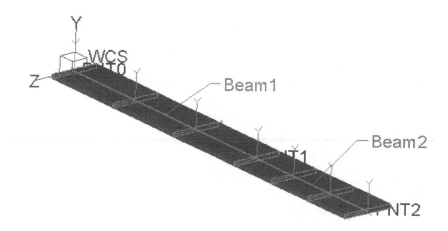

Figure 10 Model with beam elements defined

Completing the Model

The rest of the model creation is pretty routine.

Constraints

In the **Home** ribbon, select

Displacement

Name the constraint *[fixed]* (in **ConstraintSet1**). Set the **References** type to *Points*, and click on the left point on the beam. Set all the translation and rotation constraints **FIXED** for this cantilevered end[4]. Accept the dialog and move on to the other constraint:

Displacement

Name this one *[roller]* (also in **ConstraintSet1**). Set the *Points* type and pick on the middle support on the beam. Here we want to simulate a roller, so FREE the Z rotation constraint and the X translation constraint (all others FIXED). Accept the dialog.

Loads

We'll apply a single point load on the tip. Select:

Force/Moment

Name the load *[download]* (in **LoadSet1**). Set **References** to *Points*, and pick the point at the tip. Enter a load Y component of *-200* (note the units are lbf). Accept the dialog. Go to the *Simulation Display* and experiment with the settings there (for example: change the arrow setting to **Tails Touching**, turn on/off **Display Names** and **Load Value**). This completes the model, which should now appear as Figure 11.

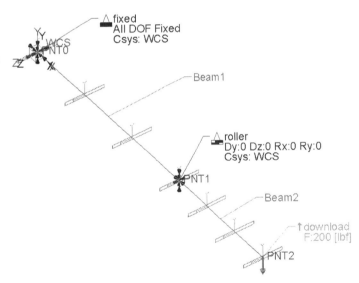

Figure 11 Diving board model completely defined

---

[4] You might wonder why we need to fix the translation in Z and rotation in X and Y. Since we will load the beam only in the Y direction, are these necessary? Based on the loading, we expect no translation or rotation in these directions. Come back later and see what happens if you free these constraints.

Analysis and Results

Performing the analysis involves the usual steps. Open the ***Analyses and Studies*** window and select

File > New Static

Enter a name *[divboard]*. Make sure constraint and load sets are selected. Set a convergence method ***QuickCheck***. What is the default polynomial order? Select the **Output** tab and change the **Plotting Grid** to **10** - this will put more points along our graphs a bit later. Then examine the ***Run Settings***. Set up the usual locations for temporary and output files. Then accept the dialog and ***Start***.

Open the **Summary** window and check for errors. Assuming there are none, change to a ***Multi-Pass Adaptive***. Set a 1% convergence criterion. You can leave the maximum polynomial order at **6**. Re-run the analysis. Delete the existing output files and open the **Summary** window. The run converges on the 2nd pass, maximum edge order 4, with zero error[5]. The maximum bending stress **max_beam_bending** is 2,670 psi, and the maximum displacement is -0.998 inch in the Y direction. How are the load magnitude, maximum deflection, and total strain energy related? Do these make sense?

Now on to the result windows.

Deformation and Bending Stress

We will create the usual result windows for stress and deformation, with one slight difference. Name the first window *[bending]*. Get the output directory **divboard**. Select a **Fringe** plot and select ***Quantity(Stress , Beam Bending)***. Under the **Display Options** tab, check the box beside **Deformed** and **Overlay Undeformed**.

Copy this window to another called *[deform]* and change the definition to produce a model displacement animation. Check the boxes to ***Show Loads*** and ***Show Constraints***.

When you ***Show*** these two windows, because you are looking at line elements only you have to be careful about the view orientation - check the XYZ coordinate triad. You are probably looking at the beam in the Creo default direction. Use the ***Saved Orientations*** toolbar button to select the **FRONT** view. The bending stress figure is not reproduced here. It shows that the maximum bending stress occurs at the roller support. The deformed shape of the diving board is shown in Figure 12. Note that the slope at the left end is zero, as it should be for a cantilevered support. What is the deformation scale?

---

[5] The first pass (and the Quickcheck for that matter) were done with 3rd order elements. For simple beams with point loads, the exact solution for displacement is typically quadratic. This explains the very fast convergence. The Quickcheck solution was actually right on, but Simulate had to do one more pass in the MPA to verify convergence.

Figure 12 Deformed shape of the diving board

Shear Force and Moment Diagrams

Now for some new forms of result windows: the shear force and bending moment diagrams. *All forces and moments are computed relative to the BACS.* Copy one of the existing windows to a new one called *[sfbm]*. Change the display type to *Graph*, and the **Quantity** to *Shear & Moment*. Deselect all options except *Vy* and *Mz* as shown in Figure 13. Leave the graph location set to **Beams**, then pick the *Select* button and, starting at the left end, click (with Ctrl) on each beam element. As you pick elements they will highlight in green. When you have picked both elements, middle click. Note that the highlighted element end point will be on the left end of the graph. You can toggle this to the other end if desired; for now leave it at the left end. Select *OK*.

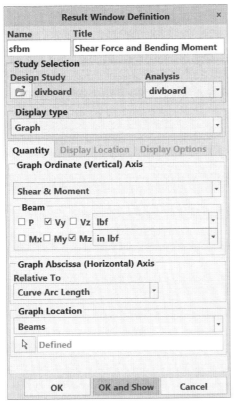

Figure 13 Result window definition for shear and bending moment diagrams

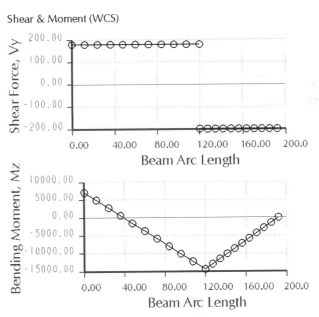

Figure 14 Shear force and bending moment diagrams

Select *OK and Show*. We get the combined shear and bending moment diagrams for the two beam elements shown in Figure 14. To change the layout of these graphs, use the *Edit* command in the **Format** ribbon.

IMPORTANT POINT: From your knowledge of simple beam theory, what sign convention does Simulate use for shear and bending moment?

As a check on these results, the shear force should be constant on each element (and 200 on the right element). The bending moment should (at least) vary linearly on the right element, be zero at the right end, be a maximum at the roller, and non-zero at the left end. Qualitatively, this diagram looks fine.

As an exercise later, find out what happens to these graphs if we reverse the direction of the BACS Y-axis (set the beam section orientation vector to <0, -1, 0>).

Changing the Constraint

We'll change the constraint on the left end to a pinned joint instead of cantilever. Find the constraint **fixed** in the model tree. Select it, and in the mini toolbar, select

Edit Definition

Change the Z rotation constraint from fixed to **FREE**. Run the Multi-Pass Adaptive analysis again. The maximum displacement in the Y direction is now -1.17 inch, so it has increased a bit from the previous case - the diving board is a bit less stiff as expected. The maximum bending stress is the same (Why?). Can you explain the value of the total strain energy?

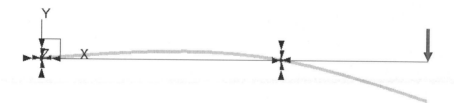

Figure 15 Deformed shape of the diving board with pinned left end (Z rotation FREE)

The deformation change from the previous case is not very pronounced. With a pinned end, the beam is free to rotate and should have a non-zero slope at the left end. You might like to increase the scale for the deformation to see the differences from the previous constraint case a bit clearer.

The new shear and bending moment diagrams are shown in Figure 16. Note that the bending moment goes to zero at the left support, as it should.

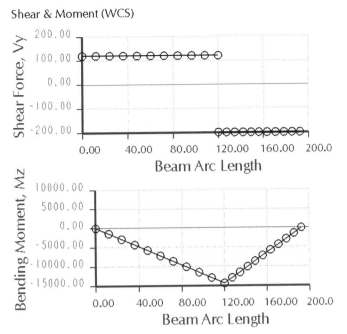

Figure 16 Shear force and bending moment
diagrams for beam with pinned left end

Now on to the second example, which is a bit more complicated. You might like to take a stretch break here.

Example #2 - Distributed Loads, Beam Releases

This example is a bit more complicated and will introduce the use of distributed loads and beam releases.

The Model

A drawing of the model is shown in Figure 17. We will use SI units: lengths in meters, force in Newtons. The steel beam cross section is an I-beam. To accommodate the distributed loads, and to provide sites for beam releases in the second part of the example, the model is divided into 4 beam elements. The origin of the WCS XY system is at the left end.

IMPORTANT NOTE: This beam model violates the Simulate guidelines for use of beam elements. This guideline is that the *ratio of a beam element length to its largest cross section dimension (its aspect ratio) should be greater than 10:1*, that is, the beam element should be long and slender. This is a normal assumption even for simple beam theory. In short, stubby beam elements, shear takes on an important role not accounted for in long slender beams. We are using this beam here strictly for demonstration purposes.

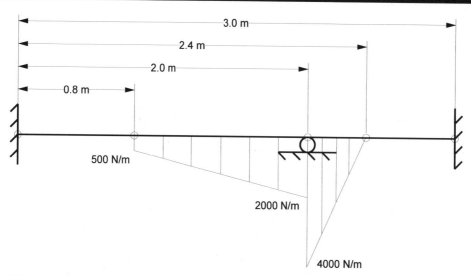

Figure 17 Indeterminate beam with distributed loads (dimensions in meters)

Beam Geometry

Start a new part in Creo called *[cbeam]*. Use the default part template, but change your units to the MKS system (meter - kilogram - second). The beam will lie along the X-axis of the default coordinate system. Create 5 datum points at the X locations (0, 0.8, 2.0, 2.4, and 3.0). These will be numbered PNT0 through PNT4.

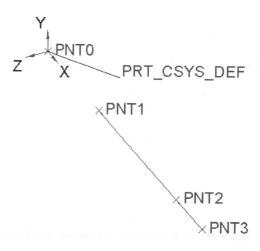

In Simulate, distributed loads can only be applied to curves (not directly to beam elements). Therefore, we must create a couple of datum curves, either here or when we move into Simulate. These are between points PNT1 and PNT2, and between PNT2 and PNT3.

Figure 18 Creo model of beam (datum points plus two datum curves)

See Figure 18. Use the *Datum Curve* command, and the *Curve through Points* option. You must create two separate curve features.

With the Creo model completed, select

> *Applications > Simulate*

and proceed with the default model type (3D).

Defining Beam Elements

We will create four beam elements. The end two use the datum points, the middle two use the datum curves. In the **Refine Model** ribbon, select

> ***Beam***

The first beam element name is **beam1**. The **References** type is **Point-Point**. Pick the points **PNT0** and **PNT1**. Select the *More* button in the **Material** area. Select **STEEL** and add it to the model. The **Y Direction** is defined using the default **Vector** (0,1,0). Select the *More* button beside **Beam Section**. Then select *New* and name the section *[ibeam]*. Select **Type(I-Beam)** and enter the following dimensions (notice that these are in meters):

flange width	b	*0.1*
flange thickness	t	*0.015*
web height	di	*0.10*
web thickness	tw	*0.01*

Select the *Review* button to see all section properties listed in the Browser (where you can easily get hard copy).

Accept all the dialog windows. The shaded element appears showing the section icons. You might like to spin the view to see these clearly.

Select *Beam* again. The next element is named **beam2**. Accept the **References** default type (*Edge/Curve*) and click on the datum curve between PNT1 and PNT2. It highlights in green, and a magenta direction arrow is shown giving the direction of the BACS X-axis. If you click on the arrow, you can reverse its direction. Leave it pointing in the positive WCS X direction. Select *Steel* in the **Material** pull-down list. The **Y Direction** uses the default direction (0, 1, 0). The rest of the data is already filled in. Accept the beam element.

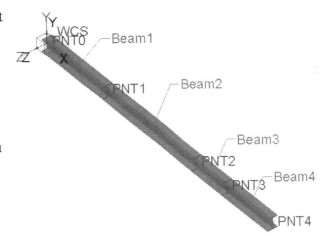

Figure 19 Beam elements created

Create the third element, **beam3**, on the datum curve between PNT2 and PNT3 in the same way as beam2. This should take just five mouse clicks. Finally, create the fourth beam element, **beam4**, using Point-Point and picking the last two datum points PNT3 and PNT4. All the elements should appear as Figure 19.

If the screen is going to get a bit cluttered up, use the ***Simulation Display*** command and window to set up the model to your liking.

Open the ***Idealizations*** entry in the model tree to see the beam elements listed there.

Completing the Model

Constraints

The beam left and right ends are fixed (cantilever) and the middle support is a roller. These are all point constraints, with the difference being whether we allow rotation about the WCS Z axis. In the Home ribbon, select

> ***Displacement***

Name the constraint *[fixed]* (in ConstraintSet1). Change the **References** type to *Points*. Select the far left and right points. Set all the degrees of freedom as fixed. ***Accept*** the dialog.

Now for the point in the middle. Create a final displacement constraint here called *[midspan]* (in ConstraintSet1). Select on the middle point (PNT2). This is the roller support so free only the X translation and Z rotation.

Distributed Loads

Simulate can define distributed loads only on curves (which is why we made them!). The load is transferred to the beam element(s) created on that curve. The load distribution can be set up using either built in functions (linear, quadratic, cubic, quartic) or specially defined user functions. We have two simple linear distributed loads in our model (recall Figure 17). The distributed load on the model is obtained as a combination between a shape function and a magnitude. We will see two variations for how these can be used.

In the **Home** ribbon, select

> ***Force/Moment Load***

Call the load *[linload1]* (in LoadSet1). Set the **References** type to *Edges/Curves*, and pick the datum curve between PNT1 and PNT2.

Now pick the ***Advanced*** button. In the **Distribution** pull-down list pick ***Force Per Unit Length*** and in the **Spatial Variation** list pick ***Function of Coordinates.*** Just below this, pick the button *f(x)* (watch for the pop-up hint *List Available Functions*). In the **Functions** window, select *New*.

In the next window (Figure 20), enter a name *[linear1]* and description. Note the default coordinate system. Change the **Definition Type** to *Table*. Pick the *Add Row* button at the right and specify the table to contain 2 rows. Now enter the data shown in Figure 20 - compare this to the load we are creating as shown in Figure 17. Select the *Review* button. There will be a warning message about units. Then select *Graph* to see a plot of the load function. Return to the window in Figure 20, then select *OK* (twice) to return to the **Force/Moment** window. So far, we have created the distribution shape, but not actually applied it, or specified its units.

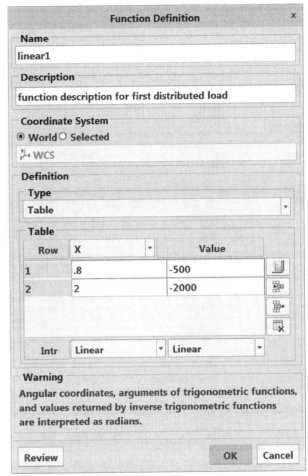

Figure 20 Defining the functional shape for a distributed load

In the Force/Moment window (Figure 21), enter a magnitude multiplier of **1.0** for force in the Y direction. Note that this is where we specify the units for the load. *Preview* the load to check the functional variation and direction. See Figure 22. Notice how the function shape (Figure 20) and load value (Figure 21) combine to create the desired load. If all is well, select *OK* and proceed to the next load. When you accept the definition, the load icon will change.

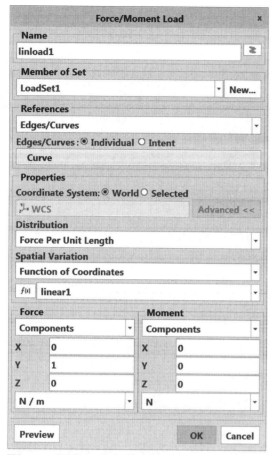

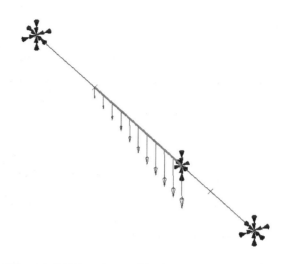

Figure 22 Preview of linearly distributed load on beam

Figure 21 Load definition window for distributed load #1

We'll create another distributed load on the next element along the beam using a somewhat different definition style. Start with the *Force/Moment* button. Name this one *[linload2]* (in LoadSet1). Set the **References** type to *Edges/Curves* and pick the next datum curve. Once again select the **Advanced** button and use **Force Per Unit Length** and **Function of Coordinates**. Click the *f(x)* button to list available functions, then *Duplicate* the definition of **linear1** to a new definition. *Edit* the new function and call it **linear2**. Instead of directly entering load values, we will define only the shape here. In the table, enter the following data:

| 1 | 2.0 | 1 |
| 2 | 2.4 | 0 |

This sets up the desired shape of the function (note that the X column is giving the locations relative to the WCS), with a unit maximum magnitude. Accept the definition and back in the **Force/Moment** window enter a magnitude in the Y direction of -4000 (units: N/m). To create the actual applied load, Simulate multiplies the magnitude times the shape function. This second method is perhaps more convenient since the load magnitude and direction are controlled directly from the Force/Moment window, instead of burrowing into the shape function. Furthermore, the magnitude can also be driven by a model parameter (check the RMB pop-up menu). Even further, we could use the distributed load in a superposition with other load sets, use a unit load magnitude, and scale the load in the result window. *Preview* the load, and if it is satisfactory, select *OK*.

Change the **Simulation Display** settings to **Tails Touching** and **Arrows Scaled**. The display will show more or less the relative size of the loads. The complete model is shown in Figure 23. Spend some time exploring the model tree to see how the features of the model are represented there.

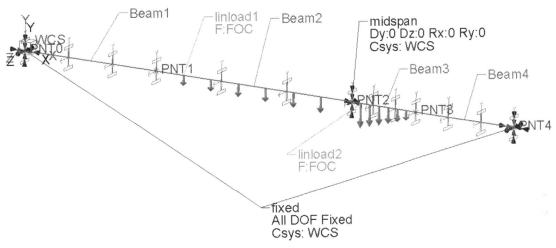

Figure 23 Completed beam model

Analysis and Results

In the **Home** ribbon, select

Analyses and Studies > File > New Static

Enter the name *[cbeam]*. The constraint and load sets (**ConstraintSet1** and **LoadSet1**) should be highlighted already. Select a *Quick Check* and using the **Output** tab, change the **Plotting Grid** to 10 - this will give us smoother curves in the result windows. Accept the analysis definition, check your *Run Settings*, and *Start* the analysis. Open the **Summary** window and look for errors. Note the resultant load on the model (-2300 N) - this checks. Assuming there are no errors, set up a *Multi-Pass Analysis* with a 1% convergence. Leave the maximum edge order at 6. *Run* the new analysis. In the **Summary** window you will see that convergence is obtained on pass 3, with not exactly a zero error (Why? What order element would be required to give an exact deformation solution for a linearly varying load?). The maximum displacement in the Y direction is -2.35e-5 m (0.0235 mm), and the maximum bending stress is 1.34E6 (1.34 MPa).

Result Windows

We will create windows showing the deformation of the beam, and the shear force and bending moment diagrams.

Create a result window "***Deformation***" and set up an animation of the displacement. The maximum deformation will look like Figure 24 (what is your view orientation and scale?). As usual, compare this with the anticipated result, and pay close attention to the constraints. The slope of the deformed beam looks like it is zero at each cantilevered end, and the deflection at the middle support is zero, as expected.

Figure 24 Beam deformation (note scale factor!)

Now set up a new result window *[sfbm]* to show the shear force and bending moment diagrams. Set up an appropriate title, select ***Display Type(Graph)***, ***Quantity(Shear and Moment)***, check *Vy* and *Mz*. Select ***Graph Location(Beams)*** and use the selection button to pick the four beam elements. Pay close attention to which end of the beam will be placed at the left side of the graphs. Select ***OK andShow***. The result window graphs are shown in Figure 25, which has been reformatted for presentation here. We can look for things like continuity of bending moment along the beam, bending moment in the beam at each end, discontinuity in shear at the middle support, and so on.

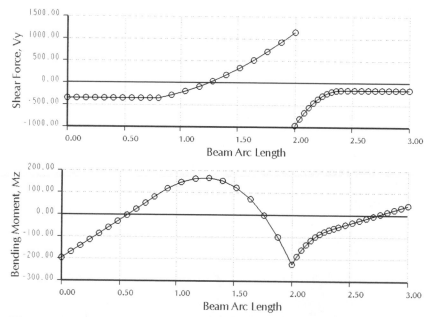

Figure 25 Shear force (top) and bending moment (bottom) diagrams for the I-beam model #2

Note the continuity (smoothness of curvature) of the deformed shape in Figure 24. In the shear and bending moments in Figure 25 note the non-zero bending moments at the connections between the first and second element (X = 0.8 m), and between the third and fourth element (X = 2.4 m). You can easily obtain the shear force and bending moment data in a number of formats using *File > Save As*. Close the results windows.

Beam Releases

A beam release is used to change the type of connection between adjacent beam elements. For a default (unreleased) connection, all six components of force and moment at the end of one element are carried through the connection to the next element. This results in continuity of these internal forces/moments along the beam (except at constraints or point loads). In this section, we will look at how we can interrupt this continuity using beam releases. In the present example, we will modify the model so that the beam is hinged at the second and fourth points. A hinge parallel to the Z axis of the beam element means that no bending moment M_Z can be transmitted through the connection. For static equilibrium of the beam on either side of the hinge, this requires that the bending moment be zero at the hinge.

Setting Releases

We will release two points. For the first one, open the model tree and select the second beam element. The element highlights in green. In the mini toolbar, select *Edit Definition* and the **Beam Definition** window opens. The element starts at PNT1 and ends at PNT2. Near the bottom of the window select the **Start** tab. (Note that we could also choose the **End** of the first beam element.) Beside **Release**, select *More > New*. Enter a name *[releaseRZ]*. Select the *Rz* button at the bottom to free the rotation in the Z direction. Note that these are relative to the BACS (which in this model is parallel to the WCS). Accept the dialog and the beam definition. An icon close to the released end of the element (at PNT1) is displayed to indicate the release[6]. This consists of a triad with a small circle around an axis in the direction of the rotational release. For translational releases, the associated leg on the triad displays an arrowhead.

IMPORTANT: Note that we do not need to release the connecting element.

Open the model tree and expand the **Idealizations > Beams** entries. Right click on **beam2** and select *Information > Simulation Entity*. In the information window that opens, you should see **releaseRZ** listed for the end of the element. Close the window. The beam release definition is also listed in the model tree under **Properties**. You can also find the release in the Beam Definition window which can be brought up by right clicking on the element in the model tree and selecting *Edit Definition*.

Click on the fourth beam element (between PNT3 and PNT4). In the mini toolbar, select *Edit Definition*. The end to release is at the **Start** of the beam. In the **Release** pull-down list, select **releaseRZ**. Accept the dialog and the beam definition. Observe the release icon on the element.

---

[6] If this does not appear, check the *Modeling Entities* settings in *Simulation Display*.

Results with Beam Releases

With the new beam releases, rerun the multi-pass adaptive analysis. Delete the existing output files. Open the **Summary** window and note that maximum bending stress is now 2.10 MPa and the maximum Y displacement is -3.81e-5m (-0.0381 mm). Compare these with the values obtained without the releases. The deformed shape of the beam with releases is shown in Figure 26. Note the abrupt changes in slope at the points of the beam releases which mean the beam elements have rotated freely around the released joint.

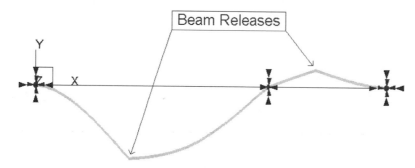

Figure 26 Deformation with beam releases

Now set up and display the shear and bending moment diagrams. See Figure 27. Observe that the shear Vy is non-zero and continuous across these released connections, and the moment has indeed become zero at the release locations.

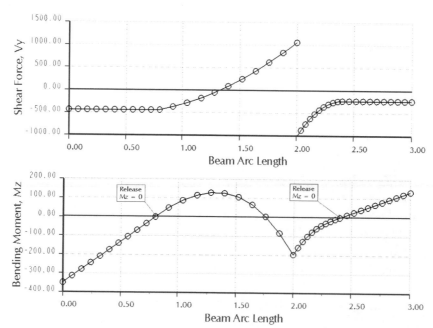

Figure 27 Shear force (top) and bending moment (bottom) for beam with releases

Some other cases where beam releases will come in handy are in modeling trusses (no moment of any kind transmitted through a connection), an expansion joint (no axial load transmitted), or a connection like a dovetail (all forces and moments transmitted except shear in one direction).

Note that in both of our beam examples, we have used loading only in the XY plane. This was for simplicity only, and is not a restriction in Simulate. We can apply loads in any direction (a situation called *skew bending*), including applied moments.

Close this model and take a break!

Example #3 - Frames

Model A - 2D Frame

The two previous models have been relatively simple one dimensional beams. Beams, of course, can be combined to form complex 2D and 3D structures. In this section, we will investigate how to create frames based on the geometry shown in Figure 28. The material is steel and the beam section is a hollow circular pipe with an outside diameter of 3" and a wall thickness of 0.25". This symmetrical shape will make specification of the beam orientation a bit easier. We will start off with a 2D frame, then move on to a full 3D frame.

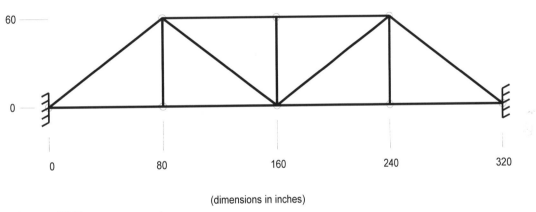

(dimensions in inches)

Figure 28 Frame geometry

Model Geometry

The easiest way to create this model is using a sketched datum curve that contains all the geometry of Figure 28. Start a new part *[frame2D]* using the default template. Change your units to inch-pound-sec (IPS). Create a sketched curve on the FRONT datum (see Figure 29). Using the sketching constraints (not shown in Figure 29), you should only need two dimensions for this. Note that we don't create any datum points.

With the geometry defined, transfer into Simulate (default model type) with

Applications > Simulate

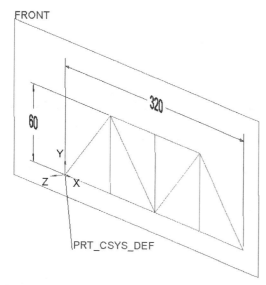

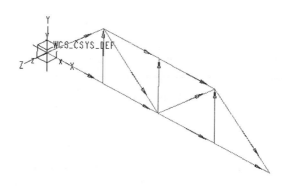

Figure 30 Selecting beam elements (arrow shows BACS X-axis direction)

Figure 29 Sketched datum curve

Beam Elements

Create the beam elements directly from the datum curve. In the **Refine Model** ribbon, select

> *Beam*

Leave the name as **beam1**. Under References, select **Edge/Curve** in the pull-down list. Separately select each of the curve segments. HINT: While holding down the CTRL key, use a right click for pre-selection of the individual segments, then CTRL-left click to add each to the selection set. The curve highlights in green and a magenta direction arrow appears on the member. This is the X-axis direction of the BACS. When all the individual curves have been selected, expand the **Beam Definition** window so that the **References** collector shows all the curves. There should be 13 listed. As you move the cursor over the list, each curve will highlight on the model. If you need to, you can reverse the X-direction by left clicking on the magenta arrow.

Beside **Material** select *More*. Transfer the material **STEEL** into the model and accept the dialog.

For the **Y Direction** select **Y-Direction defined by Vector in WCS** and change the direction to (**0, 0, 1**). This puts the BACS Y-axis for all beams in the WCS Z direction.

Beside **Beam Section**, select *More > New*. Enter a name *[hcirc]* and a description ("hollow circle"). Beside **Type**, select **Hollow Circle** from the pull-down list. Enter an outer radius (R) of **1.5** and an inner radius (Ri) of **1.25**. Select *OK* here and again in the beam definition window. All beam elements are now defined as shown in Figure 31. Notice the orientation of the BACS Y-direction. Set up a *Shaded* element display and zoom the display into one of the junctions on the top of the frame (right side of Figure 31) and you will see that this clearly is not a solid geometry model!

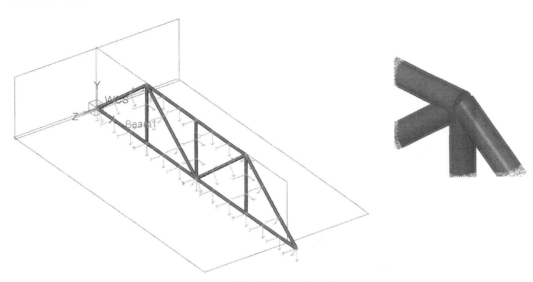

Figure 31 Beam elements created.(right) Joint detail.

Completing the Model

Constraints

We will constrain the two end points of the model. In the Home ribbon select

Displacement

Name the first constraint *[fixed_left]*. Set the **References** type to *Points*. Pick on the end of the model at the origin (left end). Observe the preselection filter and pop-up. Back in the constraint dialog window, leave all constraints (including rotations) as **FIXED** and accept the dialog.

Create another point constraint at the other end of the frame in the same way. Name the constraint *[fixed_right]*. Leave all these constraints **FIXED**. Although the constraints are the same (all fixed), we have kept this one separate so that you can experiment with it later.

Loads

We will create two load sets. The first contains a uniform load (NOTE: as in the previous beam model, this must be applied to a curve, not a beam element). The second will be a load due to gravity. Keeping these in separate load sets means we can examine their effects separately using superposition.

Start with the uniform load. Select the *Force/Moment* command and set up a new load set called *[loads]*. Name the first load in the set **[download]**. Set the **References** type to *Edges/Curves* and pick on the third element from the left across the bottom of the frame (use CTRL-RMB). Set a **Total**, **Uniform** load with a component **-1000** in the Y direction (note this is WCS). *Preview* the load and select *OK*.

Now to apply the other load. In the **Home** ribbon select

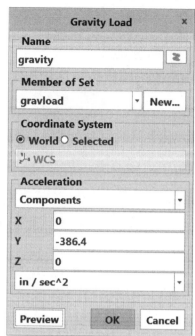

Gravity

A new dialog window opens where we define the gravity load. Name it [**gravity**] and create a new load set **gravload**. The acceleration is **-386.4** in the Y direction (remember we are using inches). See Figure 32. Select **OK**. A green icon appears at the origin (this may be obscured by the WCS and constraint symbol there)[7].

The model is now complete. See Figure 33. You can move the entity labels by selecting and using ***Move Tag*** in the RMB pop-up. Beam element colors can be changed using the button at the top right of the **Beam Definition** window (Figure 8).

You might as well delete any empty load set(s) in the model tree.

Figure 32 Defining a gravity load (IPS units)

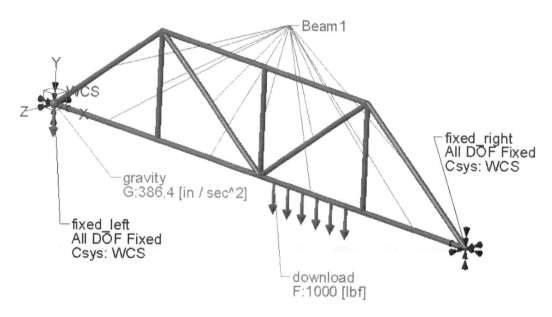

Figure 33 Model completed

Analysis and Results

Set up a **Quick Check** analysis called *[frame2D]*. The constraint and both load sets (**loads** and **gravload**) should be checked. ***Run*** the analysis (don't forget to check the ***Settings***). Check the **Summary** for errors and warnings (you can ignore a warning about

---

[7] Of course gravity acts on every element, not just the corner!

the error estimates). If all goes well, change the analysis to a **Multi-Pass Adaptive**, 1% convergence. Set a Plotting Grid of *10*, and rerun the analysis. Open the **Summary** window. The run converges in 3 passes. You might note the data for the resultant loads on the model. The resultant load for **loads** is -1000 in the Y direction. The resultant gravity load is -648, also in the Y direction, which is the weight of the frame. Note the maximum stresses and deflections for the two load sets. There is no torsion on any elements. The stresses due to the applied loads are many times greater than those due to gravity. The displacements in the Y direction are -0.043 and -0.00488 inches for the applied load and the gravity load, respectively.

Create three result windows showing displacement animations for the separate loads and for a combined load. Set the deformation scale to 1500 in each window. These are shown in Figures 34 through 36. Note the continuity of slope of each beam through each connection. Each beam shows some bending.

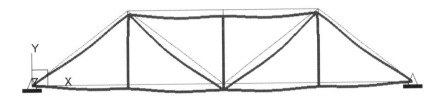

Displacement Mag (WCS)
(in)
Deformed
Max Disp 5.0628E-03
Scale 1.5000E+03
Loadset:gravload : FRAME2D

Figure 34 Deformation (gravity only)
Scale = 1500

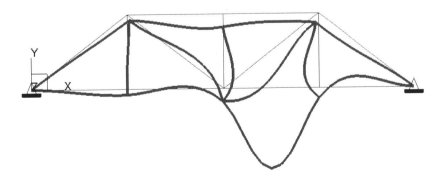

Displacement Mag (WCS)
(in)
Deformed
Max Disp 4.3317E-02
Scale 1.5000E+03
Loadset:Loads : FRAME2D

Figure 35 Deformation (load only)
Scale = 1500

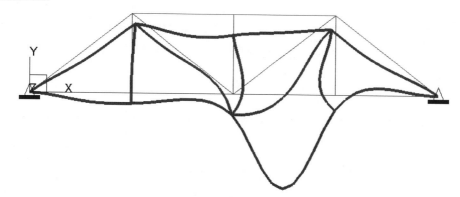

Displacement Mag (WCS)
(in)
Deformed
Max Disp 4.7291E-02
Scale 1.5000E+03
Combination

Figure 36 Deformation (combined load + gravity) Scale = 1500

Create similar result windows to show fringe plots of the **Total Von Mises** stress for each load set separately and for the combined loads. When you show all these simultaneously, you may want to set up the same legend scale in the three result windows. (Hint: check out the *Tie* command in the **Legend** group of the **Format** ribbon!) These figures are not reproduced here.

Model B - 3D Frame

We'll take the existing 2D frame and copy it to form another side of a 3D frame, then create some beam elements to connect the two frames. This is all done in Creo Parametric. In Simulate, we are eventually going to do something a bit different here: apply a displacement constraint to model a slumping foundation below one of the frame supports.

Modifying the Model

If you are in Simulate, return to Creo with *Close*, otherwise bring up the frame part in Creo. To keep our models separate, do a *Save A Copy* and save the part using a new name, like *[frame3d]*. Erase the original part file and load the new one. To create the 3D frame, we will add to the original one.

Select the sketch feature, then *Copy > Paste Special*. Leave **Dependent Copy** checked, and add a check beside **Apply move/rotate transformation**, then *OK*. Pick on the FRONT datum, then drag the small white drag handle to a distance of **80** away from the datum. Accept the copied feature[8].

---

[8] We did this *Copy* in Creo Parametric. Come back later to try out the *Copy* command in Creo Simulate.

Now create some datum curves that connect the two frames. We could go into Simulate and create these as point-point beams, but that would not allow us to apply a distributed load on the cross beams. Instead, these will be in two new sketched features. Create the first set of cross members (the lower ones) as a Sketched curve. Use the TOP datum plane for the sketching plane. In Sketcher, add additional references (for Intent Manager) by selecting existing curves along the lower edge of both of the 2D frames. Then sketch five line segments that span the gap between the frames. No dimensions should be required for this sketch, since it is referenced completely to existing geometry.

Repeat for the upper cross members. This time, the sketching plane is a make datum which is ***Through*** the top datum curves of each frame and parallel to TOP. Specify additional references for Intent Manager at the top vertices on both frames. Create the three curves in the feature to connect the top edges of the two frames.

The completed model at this point is shown in Figure 37. Note that this is in perspective view: in the **View** ribbon, select

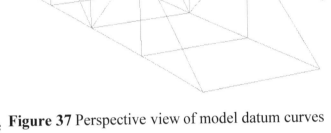

Perspective View

We are ready for Simulate. Transfer there with

Applications > Simulate

This brings the model in with all the previous elements, loads, and constraints defined on our original geometry.

Figure 37 Perspective view of model datum curves

Creating Beam Elements

We need to create elements on all the new datum curves. For the new side frame, this can be done by editing the previous element definition. Pick on any of the original beam elements and in the mini toolbar, select ***Edit Definition***. The **Beam Definition** window opens. With the **References** collector active, keeping the CTRL key pressed and alternating RMB and LMB, pick all the curve segments on the copied part of the frame (not the cross pieces). It is not critical that the direction arrows on the elements match the original. When these are selected (there should be 26 in all), middle click or select OK in the **Beam Definition** window.

For the cross pieces, we will use the same section, but the orientation is different. In the **Refine Model** ribbon, select ***Beam***. Call the beam **beam2**. The Reference is ***Edge/Curve***. Pick all eight cross members (it is easier to see these members when you are looking almost straight down on the model). These should all have the magenta arrows pointing parallel to the WCS Z direction. We will keep the same material

(**STEEL**). The **Y Direction** for these members is defined by the **Vector in WCS (0, 1, 0)**, that is, the default. Keep the same beam section **hcirc** and accept the definition window.

The screen now displays all the beam elements. Use ***Simulation Display*** to set the display how you would like.

Completing the Model

With this model, our primary aim is to examine the effect of a settled foundation under one corner of the frame. We need to set up some additional constraints, and modify the loading a bit.

To remove the load applied in the previous model, select it in the display window, then in the mini toolbar select ***Suppress***. All we should have in the model is **gravload** (although as we have seen, it doesn't do much here). The icon for this should still be shown on the WCS origin.

Constraints

We need to constrain the new front corners of the frame. Select

> ***Displacement***

Name the first constraint *[front_left]*. Make sure this is in ConstraintSet1. Set **References** to ***Points*** and pick at the left end of the frame on the new front members. Leave this point totally **FIXED**. Repeat this procedure for the other corner, naming it *[front_right]*.

The model is now complete. See Figure 38. You can just make out the gravity arrow at the far back corner.

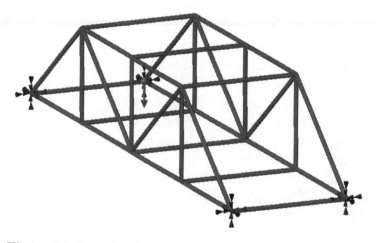

Figure 38 Completed 3D frame (Model B) with gravity load

Analysis and Results

Set up a **Quick Check** analysis called *[frame3d]*. Check your *Settings* and *Start* the analysis. Scan through the **Summary** window. There are 34 beam elements. Note the total resulting load on the model is -1687 lb in the Y direction - the total weight of the frame. Assuming there are no errors found, *Edit* the analysis to produce a **Multi-Pass Adaptive** analysis (1% convergence). Leave the maximum edge order at 6. Set the **Plotting Grid** to 10. Run the new analysis, deleting the existing

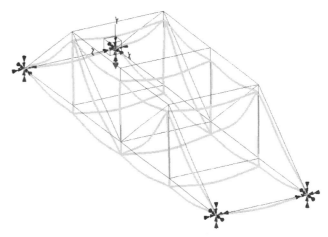

Figure 39 Deformation of Model B under gravity load

output files, and open the **Summary** window. Convergence is on the 3rd pass with a maximum edge order of 6. Note the maximum Y displacement is -0.0063 in.

Create a result window showing a deformation animation. Be sure to select the design study **frame3d**. Set a **Deformed Scale** of 2500. The deformed shape is shown in Figure 39. The animation of this looks like the frame is melting!

Forced Displacement Constraint

Now we will modify the constraint at one corner to simulate a settling foundation for the frame. This represents a forced displacement in the Y direction for the constraint.

Click on the front, right corner point constraint (or select the constraint **front_right** in the model tree). In the mini toolbar select *Edit Definition*. Change the Y translation constraint to the third button - **Prescribed**. Enter a value of *-0.1* (not very much compared to the height of the frame!). Accept the dialog.

The constraint icon does not indicate this change in the constraint. Is it in the model? Go to the model tree, expand the Loads/Constraints and ConstraintSet1 entries, right click on **front_right** and select *Information > Simulation Entity*. This information window shows the nature of the constraint. Another way of seeing this is to turn on **Value** in the **Simulation Display** settings (or just drag your mouse over the constraint).

In the static analysis study definition, you can uncheck the gravity load, then start up the previously defined static analysis. Note that with the forced displacement constraint we do not need to have any applied loads. Delete the output files and open the **Summary**. The maximum Y displacement is at the constraint, -0.10 inch (no surprise there!). Create a result window showing a fringe plot of the bending stress. In this result window, you will note that the bending stress is very low in all elements except the crossbeam entering the displaced point. Create a new deformation animation result window, or just turn on the display of the deformation in the existing bending stress. Use a scale factor of 200. The frame looks like Figure 40. Note the curvature in the crossbeam at the right end of the frame and think about the relation with the bending stress. Recall that we have set both rotation constraints to **FIXED** on all of these point constraints. You might think about this for a minute, and then find out what happens if we free these rotation constraints and run the analysis again.

Clearly, the treatment of rotational constraints is critical to the modeling of the frame (at least as far as stresses go). This is because bending stresses usually dominate in beam elements. What do you think about the way we have applied the constraints (that is, do you think they are realistic)?

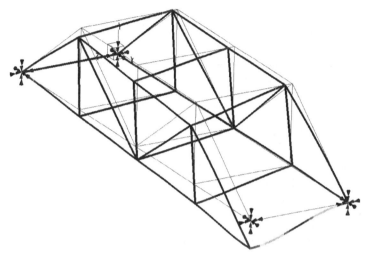

Figure 40 Deformation and bending stress fringe of Model B with displaced support on front corner

Converting a Frame to a Truss

There are two major differences between a frame and a truss structure. First, where beam elements meet in a frame they are assumed to be "welded" together so that bending moments will transfer from one element to the next. This is the Simulate default. This is particularly important at points where several elements meet - the distribution of bending moments through the joint is very complex. In contrast, a truss is composed of beams that are connected by pins (or ball joints in 3D) so that no bending moment can be transferred between elements. This assumption is often made in hand calculations since it drastically simplifies the problem. The result is that a beam element in a truss is a two-force member (it is loaded with forces only [no couples] and only at the ends) and therefore a straight element undergoes only an axial load and remains straight. There is no bending to contend with, and the computed stresses are generally much smaller.

In order to model a truss, therefore, we need to add beam releases at each end of the elements so that the rotation is not constrained and therefore no bending moment can be carried at the endpoints. We saw beam releases in the second model in this lesson. In converting a frame to a truss structure, it is not necessary to add releases to every element but only a sufficient number to ensure that bending moments are not transferred through any of the joints. This can be a very tricky business. Curiously, by allowing rotation of the element endpoint, we eliminate bending stress. This does cause some loss of stability in the structure - look at our 3D frame from the right end and imagine what would happen if there was a large side load on the top of the structure. A 3D truss would collapse without any cross-bracing.

The second difference involves the applied loading. In a classical truss structure, loads can only be applied at the pin joints where members meet. This maintains the zero-moment condition on each member by ensuring they are only loaded axially and carry no bending moment. Our 2D frame violated this loading condition since it had a distributed load along the length of one member. A true truss must be loaded with point loads applied only at the joints. While a simple truss is the easiest of structures to analyze by hand, it is the most complicated to set up for FEA due to all the simplifications required.

See the exercises for practice in converting a 2D frame to a 2D truss.

Summary

This has been a busy lesson and we have covered a lot of material. Beam models can be the simplest in terms of geometry (using just datum points and curves) but possibly the most difficult to set up in terms of modeling parameters required. The most difficult of these parameters, particularly in 3D, are related to the problem of determining beam orientation. In addition, we have not dealt with asymmetric beams, like channels or angles (with a concurrent discussion of shear center), or curved beams. Before you try that, you should consult the Simulate documentation and study the sections on the BSCS (Beam Shape Coordinate System) and the BCPCS (Beam Centroidal Principle Coordinate System). The idea of beam releases is also probably a new concept, and their use in modeling will require some additional study.

Beam elements do not need to be used in isolation. They can also be used in conjunction with solid and shell elements (in 3D). You must be careful, however, about joining the end of a beam to a solid. Recall that a beam end point has six degrees of freedom, while a solid node has only three. Therefore, transfer of a beam bending moment from a beam into the solid is a complex and advanced process.

To gain more practice with beams, you are encouraged to try some of the exercises below.

Questions for Review

1. What does BACS stand for?
2. How is the X-axis of the BACS determined?
3. How are the Y- and Z-axes of the BACS determined? That is, what is the relation of the BACS to the WCS?
4. When a load is applied to a beam, in what coordinate system are its components specified?
5. What is the BSCS? What parameters are used to define it? Relative to what?
6. What standard beam sections are available in Simulate? How can you determine their section properties (like I_{ZZ}, I_{YY}, and so on)?
7. How can you determine the direction of an element's X and Y directions?
8. In our diving board problem, how would you model the case of a person standing on one of the corners at the tip of the board (causing it to twist)? Discuss both loads and constraints for this scenario.
9. Find out if and where Simulate writes any data to a file associated with the shear and moment diagrams.
10. What is the Simulate general guideline for the beam element geometry?
11. Is it possible to create a single distributed load that spans several elements?
12. Is it possible to have two or more distributed loads acting on the same element, say in different planes (e.g. a linear distributed load in the XY plane, and a quadratic load in the XZ plane)?
13. Is it possible to have both point loads and distributed loads acting on the same element?
14. Can a point load act in the center of a beam element?
15. Is it possible to model a tapered beam? If so, how do you set it up? If not, how could you approximate the problem of a tapered beam?
16. What sign convention does Simulate use to draw shear and bending moment diagrams? Is this the same one you usually use?
17. What is the purpose of a beam release, and how is it applied? How must the beam element be created to apply a release?
18. Do you have to release *both* beam elements meeting at a point?
19. Assume there are two collinear beam elements that meet at a point. Describe the physical situation that would result in the following beam releases for one of the elements at that point: (✔ = released in the WCS direction indicated)

	Translation Released			Rotation Released		
	X	Y	Z	X	Y	Z
Case 1		✔				
Case 2					✔	✔
Case 3				✔		
Case 4				✔	✔	✔
Case 5		✔				✔

20. What is the easiest way to create a frame?
21. For our 3D frame, what is the minimum set of constraints required for static analysis?
22. What happens if you try to show a shear force and bending moment diagram of the 4 beam elements across the bottom edge of our 2D frame model?

23. What happens if you try to show a shear force and bending moment diagram of beams whose X-axes are not co-linear (for example, two elements that form an "L")?

24. In an MPA analysis of a beam model, the minimum polynomial order (used on the first pass) is 3. Why is this a good idea? (HINT: have a look at the displacement convergence graphs for our diving board model.)

25. What happens if you remove the Z translation constraints in the diving board model? What about rotation around the X or Y axes?

Exercises

1. For the beam and loading shown below, find the maximum bending stress. Plot the shear and bending moment diagrams. Redo the problem assuming pinned ends. The cross section is a hollow rectangle, 4" high and 3" wide with a wall thickness of 0.25". The material is steel. The load distribution on the right is quadratic with its apex at the support.

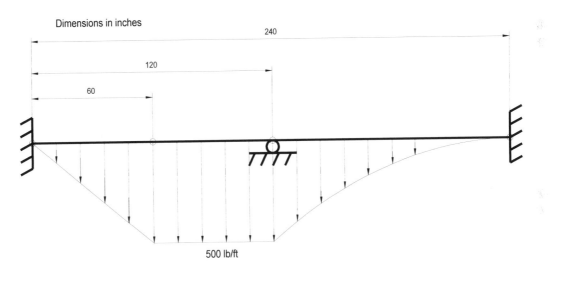

2. Redo question #1 assuming that the left end is pinned and the right end is on a roller. Compare the stress levels and deformation.

3.   A concept sketch for a transmission tower is shown below (2D model only). During a hurricane, the loads on the tower caused by the wind force on the cables are deflected 30° from the vertical. Using a frame model, find the maximum stress and deflection in the tower. The steel cross section is a hollow circle, OD 8.0 cm, with a wall thickness of 5.0 mm. You will have to estimate a couple of dimensions based on the sketch below. Do you need to consider the gravity load? Are any members in danger of buckling? (Hint: show a fringe plot of the axial load in each element.) If you are feeling ambitious, convert this to a 3D frame model, making suitable modifications so that the cable loads are at single points on each side of the tower.

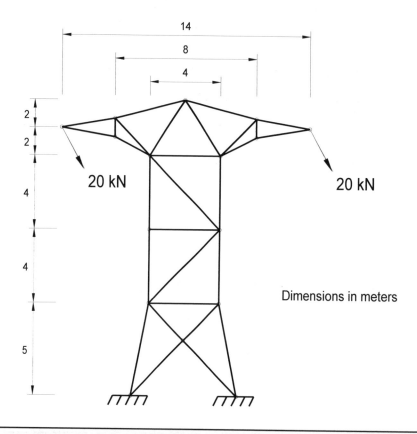

4.   Convert the frame of exercise #3 into a truss model by adding the necessary (and sufficient) beam releases. How do you modify the constraints on the ground? Compare the deformation of the truss to that of the frame. Verify that none of the elements carry any bending load or stress. Compare the stress values in the truss (what direction are these?) to those in the frame. What happens to the degrees of freedom, loads, and deflections, in the direction normal to the truss plane?

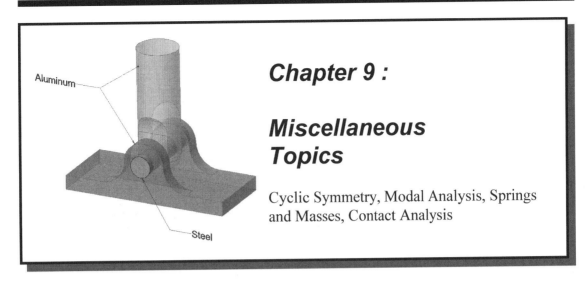

Overview of this Lesson

In this lesson, we will examine a variety of modeling tools to round out our exploration of Creo Simulate stress analysis models. As mentioned earlier, cyclic symmetry is a close relative of axisymmetry and we will look at how it is set up for a solid model. In another model, we will set up an example of the use of idealizations of linear springs and a point mass (with a gravity load). In this example, we will further explore how to create our own measures to extract precise information from the results. The same model will be used to introduce procedures for carrying out a modal analysis. This analysis will determine natural frequencies of vibration and the associated mode shapes. Finally, we will look at some variations for modeling contacting surfaces in a simple assembly.

As usual, there are some Questions for Review and a few Exercises at the end of the lesson.

Cyclic Symmetry

Cyclic symmetry is a form of geometric constraint. It is similar to axisymmetry in which the model is obtained by revolving a planar section (or curves defined on a plane) around an axis of revolution. Axisymmetric models are determined by 2D geometry and were covered in Lesson 6. In a model with cyclic symmetry, a 3D geometric shape is repeated identically several times around the axis in an equally spaced pattern. The geometry is not continuous (and therefore not axisymmetric) but cyclic. This is illustrated in Figures 1 and 2. The first figure shows a complete centrifugal fan impeller (somewhat simplified!). Because of symmetry about a horizontal plane, we can cut the fan in half, as seen in the second figure. If the loading is identical on all the blades (perhaps due to the centrifugal load), we should be able to analyze a single blade by properly isolating it and applying constraints that capture the repetitive or cyclic symmetry. We will return to this model a bit later.

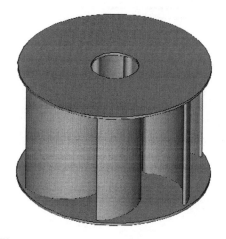

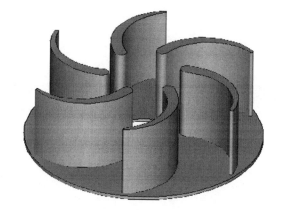

Figure 1 A centrifugal fan impeller

Figure 2 Lower half of fan impeller

The major requirement for using cyclic symmetry is that the following items are all cyclic:

♦   geometry
♦   applied loads
♦   constraints
♦   material type and orientation

Simulate will try to determine the axis for the cyclic symmetry automatically (as in the example following). If it cannot do this, you will be prompted to identify the axis yourself.

To demonstrate the procedure, we will revisit a model we dealt with earlier. This is the pressurized axisymmetric tank we treated using a 2D axisymmetric solid model. Note that the tank is fully axisymmetric, which is not required for cyclic symmetry. However, this is a simple model to create and gives us a chance to compare results with another analysis method. The completed cyclic symmetry model is shown in Figure 3. This is a 30° wedge-shaped portion of the tank.

Model Geometry

Create a new folder [**chap9**] and make it the working directory. In Creo, bring in the solid model **axitank** that we created in Lesson #6. Use *Save A Copy* to create a new copy of the part called **cycsym** in the working directory. Erase the original part and bring **cycsym** into the Creo session. The model has a 90° revolved protrusion representing one-eighth of the tank. Use *Edit* to modify the feature so that its angle is just 30°. Recall that the units for this model are IPS.

Transfer into Simulate with

Applications > Simulate

All the previous modeling entities (loads, constraints, simulation features, analyses, etc) are still there[1] - delete them all so that we are working with a 'new' model. Select these in the model tree and use RMB pop-up *Delete*.

Cyclic Constraints

To define the new constraints it will be necessary to have a cylindrical coordinate system at the origin of the datum planes. Create that now using the *Coordinate System* tool in the **Datum** group of the **Refine Model** ribbon and selecting *Type(Cylindrical).*

Create a cylindrical coordinate system as shown in Figure 3. The lower front edge of the model corresponds to the **Theta = 0** direction, and the axis of the revolved protrusion to the **Z-axis** (normal to the TOP datum). We must declare this as the current system. Go to the **Home** ribbon and in the **Setup** pull-down select

Current Coordinate System > Select

Pick on the created cylindrical system (probably **CS0**). It will turn green and should appear in a small box at the lower right corner of the graphics window.

The constraints on the two vertical faces of the model are cyclic. As our pie-shaped model is repeated (12 times) to form the upper half of the tank, the solution values on the two vertical faces must match up. Set up the cyclic constraint by opening the **Constraints** pull-down menu and selecting:

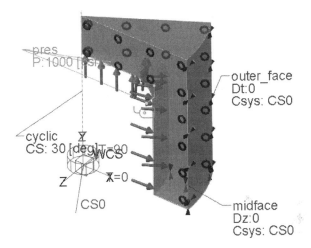

Figure 3 Completed cyclic symmetry model

Symmetry

Enter a constraint name *[cyclic]* (in ConstraintSet1). Set the **Type** to *Cyclic*. Under **References**, in the first reference collector (see the prompt for the **First Side**), select the front vertical face. Now click in the next reference collector (prompt text **Second Side**). Pick on the back vertical face. Notice that the third reference collector of this dialog window (the **Axis** area) is grayed out - Simulate was able to determine automatically where the axis is located (intersection of the two planar surfaces). Accept the dialog with *OK*. Constraint symbols (small o's) will appear on the two surfaces and the cyclic symmetry icon appears on the axis of the model as in Figure 3. Open the **Simulation**

---

[1] If you have been carefully following the lessons, the model should be set up as the solid model used in Lesson #6 with constraints for mirror symmetry.

Display window and check options for showing **Names** and **Icons** in the **Load/Constraint Display** area on the **Settings** tab.

Add a constraint *[midface]* on the lower face, also relative to the cylindrical coordinate system, to prevent translation in the **Z** direction, but free in **R** and **Theta** directions.

This model is not yet fully constrained against free body motion. What rigid body motion is still possible and consistent with the existing constraints? We need to constrain against rigid body rotation around the symmetry axis. The cyclic symmetry constraints do not do this - all they enforce is that the displacements on the cyclic planes are identical. This is why we needed the new cylindrical coordinate system. Select

> *Displacement*

Name the constraint *[outer_face]* (in ConstraintSet1). Pick on the outer curved face[2]. Confirm that the constraint is defined with respect to the cylindrical coordinate system. The constraints to set here are **FREE** for the translation in R and Z, and **FIXED** for translation in Theta. Recall that rotational constraints are ignored for solid model nodes.

Create a pressure load of 1000 psi on the inner surfaces of the tank. Finally, verify or use **Materials** to specify the material **STEEL** for the model. The model should now appear as shown in Figure 3.

Analysis and Results

Go to the **Analyses and Studies** window (you can delete the study **axitank_solid**). Set up and run a **QuickCheck** analysis called *[cyclic]*. Use ConstraintSet1 and LoadSet1 (that contains the 1000 psi pressure load). AutoGEM will create 12 solid elements. Assuming all goes well, change to a **Multi-Pass Adaptive** analysis with **10%** convergence and a maximum polynomial order 6. When you run this, open the **Summary** window. This MPA will not quite converge: as shown in the summary report, on pass 6 there are still 12 unconverged edges. We will need to modify the mesh a bit.

Change the AutoGEM **Settings** for maximum edge turn (try 30°, this will affect the fillet) and add a control for **Maximum Element Size** (try 1.0"). AutoGEM will create about 230 elements, and the MPA converges on pass 4 with a maximum edge order of 5.

The maximum Von Mises stress is now about 5050 psi and the maximum deflection $\Delta y_{max} = 0.00026$ in. Compare these to the results obtained in Lesson 6 - they are very close. Create a couple of result windows showing the convergence of the Von Mises stress and the strain energy. These are shown in Figure 4 below. Pretty good convergence here.

---

[2] You can also pick the top surface, and the inside surfaces. Come back later to add these and see if there is any effect on the model or computation. What happens if you pick the vertical surfaces with the cyclic constraints?

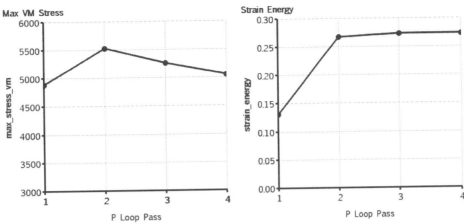

Figure 4 Convergence of maximum Von Mises stress (left) and strain energy (right) in tank with cyclic symmetry

Create result windows showing a Von Mises stress fringe plot and a displacement animation. These are shown in Figures 5 and 6. Note the stress pattern on the side (cyclic) faces are identical and very similar to the axisymmetric model in Lesson 6, and is uniform around the axis of rotation along the round. In the deformation animation, the deformation is also as expected and consistent with what we saw before. You can remove the model from your session.

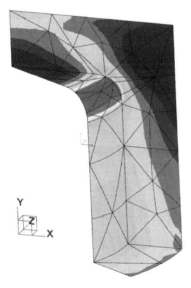

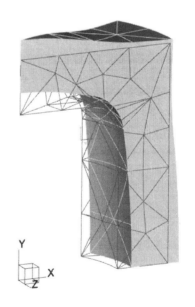

Figure 5 Von Mises stress in cyclic model

Figure 6 Deformation of cyclic model

Let's return to the example mentioned at the beginning of this section - the centrifugal fan problem. In Creo Parametric, we could isolate one of the blades using judicious cuts. Beware that the surfaces where cyclic constraints are to be applied must be identical shapes - the cyclic "instances" of the blade and end plate combination must fit perfectly around the whole fan. The included angle between the cuts must be obtained as 360° divided by a whole number, in this case, 6. This is easy to set up in Creo if you use a datum curve to define the desired shape of the cut on the end plate and then create a rotated copy (around the central axis) of the datum curve. You then create the cut using the *Use Edge* option in Sketcher and picking the datum curves.

The resulting geometry and model are shown in Figure 7. The cyclic symmetry constraints are placed on the S-shaped cut faces of the end plate. Could a shell be used for this part of the model? Depending on the geometry of the end plate cut, you may have to identify the cyclic symmetry axis for this geometry. The next constraint is due to symmetry on the top surface of the blade in the model to prevent rigid body translation along the axis. Finally, the model must be constrained against rigid body rotation around the central axis. This is easily done with a cylindrical coordinate system in the same way we handled the previous tank problem. This should probably be applied on the inner surface of the end plate. The model is loaded with a centrifugal load

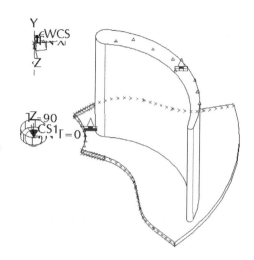

Figure 7 Model of fan blade using cyclic symmetry

(notice the symbol in the upper left corner of the figure) that appears on the axis of rotation of the fan.

Some results of running this model (326 solid elements) are shown in the figures below. Note the maximum stress levels are near the boundaries and at the junction of the blade trailing edge and the end plate, which is a reentrant corner. The maximum Von Mises stress convergence graph (not shown here) also displays evidence of a problem. We might have expected this! This model would need some re-work before providing good results.

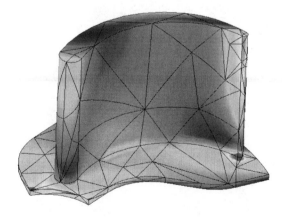

Figure 8 Von Mises stress on fan blade in model with cyclic symmetry. Note hot spot at blade root near trailing edge.

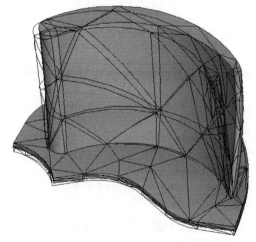

Figure 9 Deformation of fan blade and end plate

Springs and Masses

In this section we will look at two more idealizations - springs and masses. Springs can be either extension/compression springs or torsion springs that connect two points, or one point and the fixed ground. Masses can be either point masses or can be given inertial properties that will affect their rotation. Springs are commonly used to model compliant connections to the ground rather than perfectly rigid constraints. Masses are used to add inertial effects to dynamic models without forcing the creation of a lot of solid or shell elements.

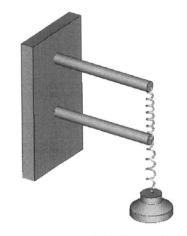

Figure 10 Model with spring and mass entities

We will examine a simple model composed of two short beams cantilevered out from a wall. The 6" long aluminum beams have a solid circular cross section (diameter 0.5"). The free ends of the beams are connected by a linear spring. Another spring supports a mass (we will model it as a point mass). The only load on the system is due to gravity. The physical system is depicted in Figure 10.

All springs in Simulate are linearly elastic. There are two kinds of springs: **Two Point** (connecting any two points in the model) and **To Ground** (connecting any single point directly to the fixed ground). *Two Point* and *To Ground* springs can come in any orientation (specified somewhat like a beam). The spring properties required are its *Stiffness* (extensional, torsional, or mixed) and its *Orientation* (for Two Point springs only). It is possible to constrain a model entirely using *To Ground* springs, rather than having "hard" constraints. You must be careful in that case that every relevant degree of freedom has some spring stiffness associated with it.

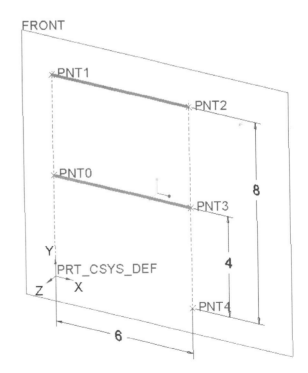

Figure 11 Creo model of system (dimensions in inches)

Masses also have a number of options. The mass element is applied at a point (which can be created on the fly). We are not restricted to point masses, though, since the mass element can be given inertial properties that would respond to rotation of the model (for example in a centrifugal loading case).

Model Geometry

Our idealized model of the beam/spring system will consist of two beam elements defined on datum curves. Start Creo Parametric and create a new part called *[twobeams]*. Set your units to IPS. On the **FRONT** datum plane, create 5 datum points with the dimensions shown in Figure 11. These are numbered **PNT0** through **PNT4**. You might like to keep **PNT4** in a separate feature since we will be deleting just this point a bit later in the lesson.

Create a couple of datum curves to connect the top pairs of datum points. The completed Creo model is shown in Figure 11.

Bring the model into Simulate with

Applications > Simulate

Accept the default model type (3D).

Creating the Elements

This model will have three kinds of idealized elements. First, we'll create the beam elements. In **Refine Model** select

Beam

Use the default name **Beam1**. In the **References** area, the default is **Edges/Curves** in the pull-down list. Pick the two datum curves. On each beam element, the magenta arrow indicates the beam's X-axis points parallel to the WCS X-axis. In the **Materials** area, select *More* and add **AL2014** to the model and select *OK*. For the **Y Direction**, use **Vector in WCS** and keep the default direction (0, 1, 0). On the **Start tab** beside the **Beam Section** pulldown area, select *More > New*. Name the section [circle]. Select *Type(Solid Circle)* and enter a radius of **0.25**. Accept all the dialogs. When you click on the background the beam section icons appear on the model in blue (or whatever color you select in the *Color* button at the top right of the **Beam Definition** window).

Now create the springs. Select

Spring

Call the first spring **spring1** (default). The default type is a *Simple* (i.e. no lateral stiffness), **Point-Point** spring. Using the two reference collectors, pick on the points at the ends of the two beams (PNT2 and PNT3 in Figure 11). Enter an extensional stiffness of **2000** (note units of lbf/in). See Figure 12. You can also obtain this stiffness from a Creo model parameter (see Exercise #2). Accept the dialog. A yellow spring icon appears on the model.

If you pick an *Advanced* spring type, things get quite a bit more complicated. Select the *More* button beside **Spring Properties** to see the options. You can set lateral and torsional stiffness, relative to a coordinate system fixed to the spring. In this case, you must also worry about the orientation of the spring. This is given in much the same way as the orientation of a beam element: the spring X-axis connects the two points and you specify the direction of the spring Y-axis in the WCS.

The last **Type** option is **To Ground**. This can be used to provide "soft" constraints to points on a model that offer resistance in proportion to displacement instead of "hard" constraints that restrict motion totally.

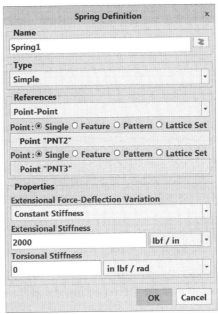

Figure 12 Defining a simple spring

Create the second simple spring between points PNT3 and PNT4. This is also a *Simple*, *Point-Point* spring. The extensional stiffness is **1000.** Accept the dialog.

Now to create the mass. In the **Refine Model** ribbon, select

Mass

Note the defaults. Call it **mass1** and pick on the lower point (**PNT4**). We want a weight of 10 lb. The mass is (weight/gravity = 10 / 386.4 =) **0.02588**, since in the IPS system g = 386.4 in/sec^2. What are the units of mass in IPS?

Note that by picking an advanced mass type, we can also specify moments of inertia, which might be important if the mass was going to rotate. Accept the dialog and a dark blue mass icon appears on the model.

Observe where all the idealizations appear in the model tree.

Loads and Constraints

We'll cantilever the beams out from the wall. In the **Home** ribbon, select:

Displacement

The constraint name is *[fixed]* (in ConstraintSet1). Set *References* to *Points* and pick (with Ctrl) the two points of the beams at the wall (**PNT0** and **PNT1**). Leave all degrees of freedom fixed (including rotations) and accept the constraint.

The gravity load is applied with

Gravity

Note that gravity is relative to the current coordinate system (WCS). Enter a name *[gravity]* in LoadSet1 and enter a value of -386.4 (in/sec²) in the Y direction.

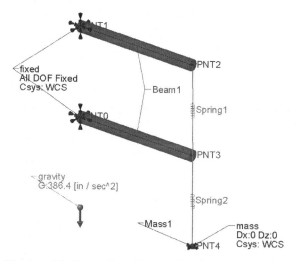

Open the ***Analyses and Studies*** window. If you select the ***Info > Check Model*** command, no errors will be found. However, if you try to run the model now, it will fail with an "insufficiently constrained" error message. Why? Since the mass is just hanging on the end of the spring, in addition to the desired vertical

Figure 13 Completed FEA model

motion it also has two rigid body degrees of freedom (translation in horizontal X and Z directions). For our purposes, we can constrain it to move only in the Y direction. Do that now (make sure the constraint is in ConstraintSet1)[3]. Apply the constraint to **PNT4** and leave all the degrees of freedom as **FIXED** except for translation in the Y direction. The completed model should look like Figure 13.

Analysis and Results

Set up a **Quick Check** static analysis. ***Run*** the analysis (you can ignore the warning about stress concentrations) and open the **Summary** window. Note the Model Summary at the top (2 springs, 1 mass, 2 beams). Assuming there are no errors, ***Edit*** the analysis to produce a **Multi-Pass Adaptive** analysis with 5% convergence, maximum polynomial order 6. Run the MPA analysis. Open the **Summary** window. The run converges on pass 2 with max edge order 4. The resultant load in the global Y direction is -10.24 lb. The extra 0.24 lb is due to the weight of the beams! Check out the stresses and displacements. What is the maximum displacement, and where does it occur? It is probably the deflection of the point mass. Create a result window to show a deformation animation. The springs and the mass are not displayed. Change the window definition to:

> ***Display Type(Fringe)***
> ***Quantity(Displacement)***
> ***Component(Y)***

and show the window. The tip displacement of the lower beam is greater (in magnitude) than that of the upper beam. What are the deflections of the two beam tips exactly? And what about the mass? Can you use Dynamic Query here? Here's how we can find that out by defining our own *measures*. Close the result window.

---

[3] Come back later and see if converting the lower spring to an Advanced type and adding Kyy and Kzz properties (remember that directions are defined relative to the spring coordinate system) solves this problem instead of adding the extra constraint.

Defining Measures

In the **Run** group of the **Home** ribbon, select

> ***Measures > New***

Call the first measure *[defy_top]*. Click the
Details button and enter a description. Select
the following in the pull-down lists:

> **Quantity (Displacement)**
> **Component (Y)**
> **Spatial Eval (At Point)**

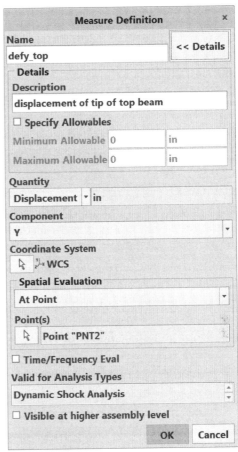

See Figure 14. Select the **Points** button and
click on the end point on the top beam. ***Accept***
the dialog with a middle click. A small icon
appears at the tip of the beam.

Copy this definition (use ***Duplicate*** in the RMB
memu) to create a new one; click on the datum
point at the end of the lower beam. In the
Measures window, select the new measure and
open it with ***Edit Definition***. Change to a new
name *[defy_mid]*. Make sure to select the Y
displacement. Keep the existing settings and
accept this definition.

Figure 14 Defining a measure

Copy **defy_mid** to a new measure *[defy_mass]*
and select PNT4.

We now have three of our own measures. Unfortunately, custom measures are not
retroactive - they are only computed during the analysis run and not after (plan ahead!).
Go to the ***Analyses and Studies*** window and ***Start*** the analysis again. Open the
Summary window. At the bottom of the Measures list in the summary we find our three
measures. They are **defy_top** = -0.0101 in, **defy_mid** = -0.0123 in, and **defy_mass** = -
0.0223 in. What is the extension of the lower spring? What is the force in the lower
spring?[4] Does this value make sense? How much of the 10 lb weight is carried by the
upper beam?

This model is the basis for an exercise at the end of the lesson, so make sure you save it.

Return to Creo for the next section of the lesson.

---

[4] Note that you can define a measure for a spring that will directly give its force.

Modal Analysis

As you probably know, when a structure is excited by a dynamic load at close to its natural frequency, you can expect trouble. Therefore, it is important to be able to predict what the natural frequency is. Furthermore, for all continuous systems, there are a number of frequencies of vibration that can occur naturally. Each frequency has associated with it a characteristic deformed shape, called the mode shape. The modes are numbered, with mode 1 having the lowest frequency (it is called the fundamental mode). We are usually concerned with the modes that have the lowest frequencies. In Simulate, we can find these frequencies and mode shapes using *modal analysis*. We will investigate this form of analysis using the beam/spring model from the previous section.

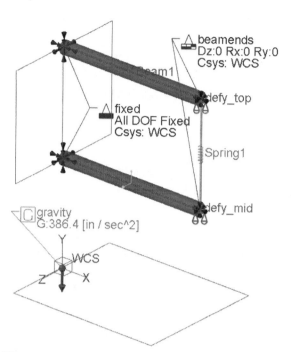

Figure 15 Model for modal analysis of beam/spring system

Using *Save A Copy*, make a copy of the **twobeams** model created in the previous section. Call the new model *[twobeams_modal]*. Erase **twobeams** from the session and bring in the new model. Delete the datum point where we put the mass (**PNT4** in Figure 11) - this also will remove associated modeling entities. Bring the modified model into Simulate.

Setting up the Model

We have no other geometry to create. However, we are interested primarily in the motion in the XY plane, so let's constrain the tips of the beams for this. Select

> *Displacement*

Name the constraint *[beamends]* (in ConstraintSet1). Set **References** to *Points* and select the two points at the ends of the beams. **FREE** the translations in X (why X?) and Y, and the rotation in Z, and **FIX** everything else. See Figure 15. (You should come back later and delete these end constraints to see what happens.)

Defining the Modal Analysis

Set up the analysis with

> *Analysis and Studies > File > New Modal*

Call the analysis *[twobeams_modal]*. Enter a description. There are quite a few differences in this window from what we have seen before. First, notice that there is no indication of the load set. For modal analysis, you don't need a load set. There are options for constrained and unconstrained analysis. Leave the default for this (consult the on-line documentation for further discussion). There are several tabs at the bottom. Under the **Modes** tab, you can select how many modes you want Simulate to find - the default is 4. See Figure 16. If you are concerned about a special frequency (for example, due to the speed of a rotating machine), you can search for modes in a near-by frequency range. Under the **Output** tab, change the **Plotting Grid** to 10. Under the **Convergence** tab, set a **Multi-Pass Adaptive** analysis with 5% convergence. Notice that this convergence is on frequency by default. Accept the dialog window.

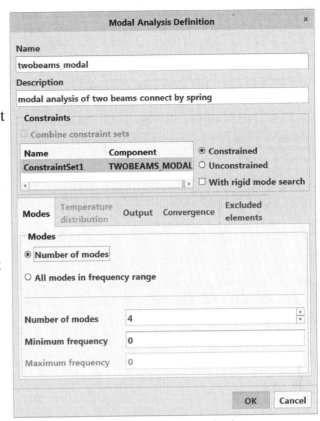

Figure 16 Defining a modal analysis

Now *Run* the analysis **twobeams_modal**. Open the *Summary* window. The run converges on pass 3 (maximum edge order 6). The four natural frequencies are:

mode #1   389.7 Hz
mode #2   1068 Hz
mode #3   2387 Hz
mode #4   2408 Hz

Showing the Mode Shapes

We'll create four windows to show the four mode shapes. Select

Results > Define Result Window

Call this first window *[m1]* and make sure you are looking at the data from the **twobeams_modal** design study. Select mode **1** with **Scaling** set to 1. Enter a title *[Mode #1]*. Set *Display Type(Model)*, and *Quantity(Displacement)*. Under **Display Options**, set up an animation with a scale of 10%. Check the options **Overlay Undeformed** and **Alternate** (this is the right button beside **Auto Start**).

Copy the current window **m1** to a second window *[m2]*. Accept the window definition as is except for selecting mode 2 and changing the title to *[Mode #2]*. Accept the edited definition.

Continue with this procedure to create windows for mode #3 and mode #4.

When all four windows are created, show them all at once. Each result window shows the mode number and frequency. Be careful about your viewing direction for these windows - observe the coordinate system triad in each window. The display should look something like Figure 17.

The first mode shows the beams moving up and down in unison (spring not stretching?). Mode 2 shows the beams moving in opposition (spring definitely stretching!). Mode 3 shows the beams moving in unison with an S-shape. All of these are in the vertical XY plane. Mode 4 shows the beams bending in opposition in a horizontal plane with the tips almost stationary (remember the tip constraints).

You might explore the effect of different constraints applied to the beam end points. For example, what happens if you **FREE** the rotations around X and Y? What happens if you remove the constraints on the tips of the beams?

Save the model and return to Creo, where you can erase it from your session.

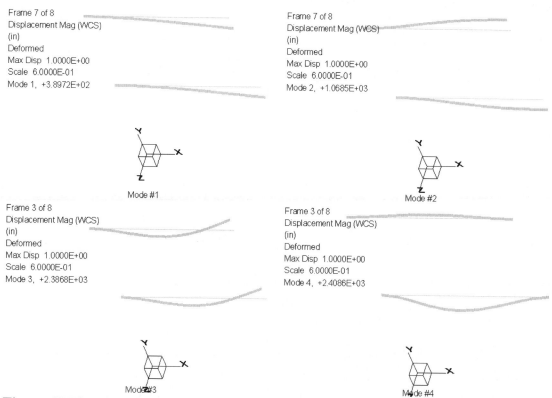

Figure 17 First four mode shapes of beam/spring system

Modeling Interfaces in Assemblies

Creo Parametric is noted for its ability to create and manage assemblies. These assemblies are easily brought into Simulate for stress analysis. Be warned that (despite what we will see here) analysis of assemblies can require very advanced modeling techniques and understanding, particularly if the assembly is a mechanism with moving parts. One of the requirements for analysis of assemblies is the proper modeling of the contact between adjacent parts. Options for how these adjacent or contacting surfaces are treated are known as *Interfaces*. There are three basic interfaces available in Simulate Structure (two additional types are available in Thermal):

- **Bonded** - the two component surfaces are essentially glued together, sharing common nodes and therefore displacement. This forms a continuous solid where mesh surfaces coincide with the component surfaces. Material properties can change abruptly across the surface. All forces are transmitted across this surface.
- **Free** - the two surfaces are not connected in any way. The surfaces can interpenetrate (even pass through each other) and no forces are transmitted across the surface. The two surfaces basically ignore each other, which would be all right if the two surfaces were only moving apart.
- **Contact** - this is the most realistic model. The surfaces cannot interpenetrate but may move apart to create a gap. Only compressive normal stresses can be transmitted across the surface (that is, the surfaces cannot "pull" on each other). Contact surfaces can be either frictionless or include friction. In the frictionless case, sufficient constraints must be supplied to the model so that the surfaces do not slip tangentially. If the surface has friction, tangential forces will be calculated by Simulate and compared to a maximum possible value determined by the normal force and a friction coefficient. If this static friction limit is reached or exceeded (the surface partially or totally slips), a warning message is created. For further information on this situation, consult the online help and look up "slippage" in the index.

We will look at the **Bonded** and **Contact** interfaces using the simple assembly shown in Figure 18. This consists of an aluminum base plate (approx. 24" X 10"), a steel pin (diameter 3", length 12"), and an aluminum connector (diameter 6"). The holes in the lugs (2" thick) on the base plate and on the connector are the same as the pin diameter. The dimensions of the lugs on the base plate and the cut out on the connector are not critical to what we are going to do - an approximate geometry here is satisfactory (although clearly your stress result values may be a bit different). When assembled, an upward force will be applied to the connector. To take advantage of symmetry in the geometry and loads, a quarter model can be created in Creo using a cut created in assembly mode. See Figure 19 (note that this model is in the positive quadrant - view is from top - left - rear; observe the coordinate systems in Figure 19).

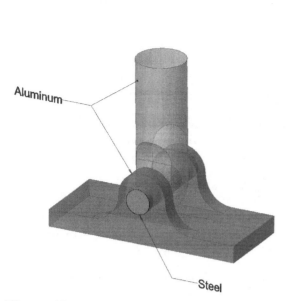

Aluminum

Steel

Figure 18 Creo Parametric assembly

ASM_DEF_CSYS

Figure 19 The FEA model

After you have created the parts, create an assembly in Creo Parametric. Make sure the units for each part and the assembly are consistent (IPS). Use a cut in the assembly to extract the quarter model (Figure 19) from the entire assembly. When the quarter model is ready, bring it in to Simulate with

Applications > Simulate

Check *Model Setup* to confirm that the default model interface type is **Bonded**. If you browse the ribbons, you may notice some new toolbar icons have appeared since we are looking at an assembly. Can you find them? They refer to model elements that only make sense in an assembly.

Using *Materials*, bring the materials **AL2014** and **STEEL** into the model. Then, select

Material Assignment

Select the two aluminum parts (they highlight in green). In the **Material Assignment** dialog window, make sure **AL2014** is selected in the **Material** pull-down list, then *OK*. Repeat this sequence to assign **STEEL** to the pin.

Constraints are applied to the two symmetry planes and the lower surface. In the **Constraints** group pull-down menu, select

Symmetry

Apply appropriate constraints (mirror symmetry) to the surfaces created with the two cutting planes arising from symmetry. Name these *[XYface]* and *[YZface]*. These

involve fixing the translation
perpendicular to the surface, and freeing
the other two translations. These are handled
automatically by the **Mirror** symmetry
constraint. Since the model will use
solid elements, the rotational constraints
don't matter. Note that surfaces of all
three parts should be included in each
constraint. Finally, apply a translation
constraint to the bottom of the base to
prevent rigid body motion perpendicular
to that surface.

Now apply an upward load:

Force/Moment

Create a load on the top surface of the
connector (**Uniform, Total Load, 5000**
lb upward). The model should appear as
shown in Figure 20.

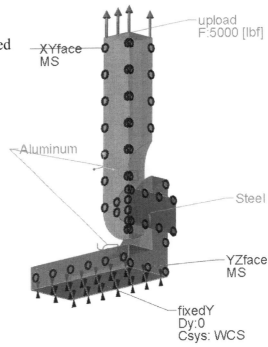

Figure 20 Model with constraints and loads
complete

Using Bonded Surfaces

If we perform an analysis of this model now, Simulate will "weld" the surface of the steel
pin to the aluminum holes. This essentially creates a continuous solid whose material
properties change abruptly as you cross the pin-hole surface interface.

Set up and run a new static analysis called *[bonded]*. Verify the model using QuickCheck
and if everything is satisfactory, rerun using an MPA convergence with 10%
convergence and maximum polynomial order 9.

Create result windows for the Von Mises fringe plot and the model deformation. These
are shown in Figures 21 and 22. Also, better check the convergence graphs. These show
a problem with the convergence of the maximum Von Mises stress.

In the stress fringe, look for the stress concentration (singularity?) in the connector where
it meets the upper surface of the steel pin. Notice the stress distribution through the lug
on the base plate and on the lower end of the connector. In the deformation, note that the
surfaces have all remained in contact with each.

Close these windows and return to the model.

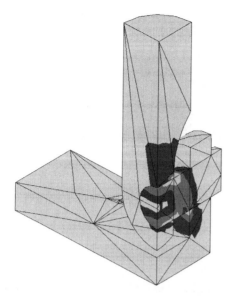

Figure 21 Von Mises stress fringes of model with bonded surfaces

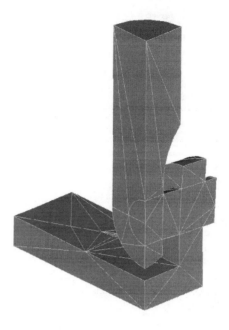

Figure 22 Deformation of model with bonded surfaces

Creating Contact Regions

We know that because of the construction of the assembly and the applied load, as the load is applied, some of these surfaces might actually separate (create a gap). This is because contacting surfaces cannot support a tensile normal stress. We can define contact surfaces that Simulate can monitor for this - the surfaces are free to move apart (normal to the surface) to cause a gap to be created but will not penetrate the other part. Contact will always give rise to compressive normal stresses. To create these contacts, in the **Refine Model** ribbon, **Connections** group, select

> *Interface*

Rename the interface as *[plate_pin]*, and change the **Type** to **Contact**. The default reference type is **Surface-Surface**, for which we explicitly identify the contacting surfaces. The other option ("**Component-Component**"), as the name suggests, allows you to just identify the components involved and Simulate will determine which are the contacting surfaces. In a complex model, as you can imagine, this is not very efficient! Pick on the outside surface of the steel pin and (using *Query Select*) the inside surface of the hole in the lug. Leave the properties at their default settings (but note the drop down list for **Friction** options here - come back later to check this out). Select *OK*. The contacting surfaces will be outlined in blue. A contact region symbol will appear along with a couple of measure icons. Some new entries have also appeared on the model tree.

Repeat the process to create a contact interface where the pin passes through the hole on the connector. The two contact region symbols are shown in Figure 23, along with some icons that indicate that measures have been defined for us. In the model tree, select each contact in turn. It will highlight on the model to confirm the surfaces are correct. If you select one of the measure icons, you can use the mini toolbar to select *Edit Definition*. This will let you get more information about that measure. There are actually four measures defined (open the model tree to verify): for each contact region, measures were automatically created for contact area and force. The icons may be overlapping on the screen. You can create other measures for the contact using the **Measures** function and selecting **Quantity(Interface)**.

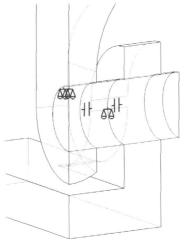

Figure 23 Contact regions and measures defined on the model

Now set up and run a new **QuickCheck** analysis called *[contact]* to see if we have any errors in the model. Turn on the **Nonlinear** checkbox, and then the checkbox for **Contacts** (actually, these should have been checked by default). Notice there is an option for Time Dependence. This is a bit of a misnomer for a static analysis, but what it means is that the load will be applied incrementally over a number of steps.

On the Output tab, we can set how many intervals we want to apply the load. The default (**Automatic Steps**) will use just two steps (time = 0 and time = 1). If you select **User-defined Output Steps** you can select how many intervals or steps you want.

For now, leave this set to **Automatic**. (Come back later and try five equally spaced intervals and compare results.) *Run* the analysis. AutoGEM creates 100 or so solid elements. Assuming no errors (you will note, but can ignore the warnings!) change the analysis to a **Multi-Pass Adaptive** analysis (10% convergence, max order 6). The run will take a bit longer and should converge on pass 4 or 5. If not, try using a denser mesh, especially in the contact regions. Since these are curved, the easiest control is to modify the maximum allowed element edge turn. In the **Summary** window, note that the contact area and force are given for each of the contact regions. There may be a warning message about mesh size and contact area. This is discussed briefly at the end of this section.

Create result windows for the Von Mises stress and a deformation animation. See Figures 24 and 25. The Von Mises fringes are very different from the bonded case (compare with Figure 21), particularly on the top of the lug, both top and bottom of the pin, and the connector. Zoom in on the contact regions in the deformation window to observe the separation of the surfaces. You may also see some interference (surfaces passing through each other). This is possibly due to our coarse mesh and high convergence tolerance. Come back later to investigate this. Contact models typically require fine tuning!

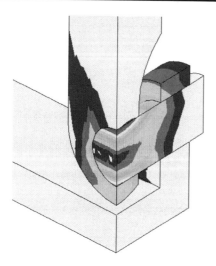

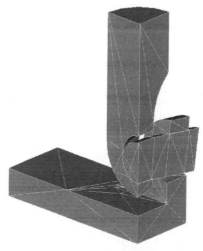

Figure 24 Von Mises stress
fringe plot with contact surfaces

Figure 25 Deformed shape
showing gaps around pin

Create another result window showing the
normal stress component in the direction of
the load (the YY component). Check the
option to show the model in the deformed
shape. The result window is shown in
Figure 26 (deformation scale is about 300).
Note the gap that has been created around
the pin, and the continuity of the normal
stress where the surfaces remain in contact.
You might like to examine the YY stress on
an XY cutting plane through the model.
Finally, create result windows to show the
convergence of the contact pressure
(**contact_max_pres**) and total contact area
(**contact_area**) for the two regions. These
are pretty well behaved.

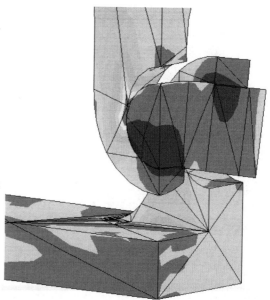

Notes for Contact Analysis

Figure 26 Normal stress (YY) fringe plot

The default contact interface is frictionless. If your model contains multiple frictionless
contact interfaces, you must ensure that components are somehow constrained in the
directions tangent to the surface. For example, the pin in our assembly was constrained
axially and against rotation around its axis by the symmetry constraints. Also, it may be
necessary to use mesh refinement so that elements in the contact zone have smaller
surfaces than the zone itself. If the elements in the contact zone are too large, you will get
a warning message in the summary report. This happened in the model here. Try to
introduce some mesh controls (like maximum element size) on the surfaces of the pin and
holes to yield a finer mesh. There is an option in the analysis definition window to
increase the mesh density on the contacting surfaces (for SPA analysis only) - beware

that this can cause a large increase in the number of elements in the model. You might also like to experiment with the number of load increments. For these variations keep track of the maximum contact force and area to see the effect of these modifications.

As mentioned earlier, you can turn on friction at the contacting surfaces. When you do this, Simulate will compare the computed force tangent to the surface (at a number of points) to the maximum possible static friction determined by the normal force component and a friction coefficient provided by you. A warning message is given if the static friction limit is exceeded at a point or for the entire surface. Note that the surfaces are not actually allowed to slip past each other - this is just a warning that there is a problem with the model (something like the Von Mises stress exceeding the material yield limit). Revisit our assembly model and turn on friction for the contact interfaces. Find out which constraints (if any) can be removed from the model, and the effect of varying the friction coefficient.

Summary

This lesson has covered a grab-bag of miscellaneous modeling topics.

Cyclic symmetry can be exploited to simplify the FEA model for appropriate cases. You must be careful that all components of the model (geometry, constraints, loads, materials, ...) are truly cyclic. Spring and mass elements will be useful in some types of models. They are easy to apply, although spring stiffness properties and orientation may take some thought. Modal analysis to determine natural frequencies and mode shapes is also quite simple. Remember that you don't need any applied loads for a modal analysis. For that matter, you do not need constraints either, in which case you will find the modes and frequencies for free vibration of the model as if it was floating in space (or resting on a very soft foundation). If there are loads applied to the model, they could affect the modal analysis (think of a vibrating wire or guitar string under varying tension - the mode shape may stay the same but the frequency changes with tension). In the analysis of assemblies, contact regions can be defined to handle the kinematic constraints between contacting surfaces.

There are some challenging questions and exercises at the end of this lesson that will require some exploration of these topics.

In the next lesson we will have a brief look at something totally different - thermal modeling in Simulate. This will include steady state and transient models, and applying a thermal load to compute thermally induced stresses.

Questions for Review

1. When can you use cyclic symmetry?
2. Find out if there are restrictions on the orientation of the cyclic symmetry axis.
3. What are the two kinds of springs?
4. What properties are required to define a spring connection between two different points on the model?
5. What properties are required to define a spring connection between a point on the model and the ground? Where/how is the ground connection located?
6. To which of the following can cyclic symmetry constraints be applied: points, edges, surfaces.
7. Is it necessary to apply a radial constraint in a model with cyclic symmetry?
8. Is it possible to define a spring whose stiffness varies with its extension? What is this kind of spring called?
9. In general, what happens to the spacing between natural frequencies as they increase?
10. What do *Free*, *Bonded*, and *Contact* mean in regards to the model setup?
11. How could you determine the size of a gap formed between two contact surfaces?
12. What happens if you accidentally specify two surfaces as a contact pair that never actually touch each other?
13. In the IPS system, what are the units of mass? Inertia?
14. How would you interpret the torsional stiffness of a spring connecting two points?
15. What happens to mating surfaces in an assembly if they are not designated as contact surfaces?
16. What happens if you try to form a pair of contact surfaces from concentric cylinders of different diameter?
17. Does a contact pair of planar surfaces allow friction between the surfaces?
18. How does the option work for automatically detecting pairs of contact surfaces? See the command *Interface > Detect Contacts* in the Refine Model ribbon.

Exercises

1.   Find out if it is possible, in an assembly, to create a spring of zero length by selecting points (on different components) that happen to be coincident.

2.   Create a model consisting of a point mass (1 kg) supported by a single spring (100 N/mm). What is the natural frequency of this simple system? What is the theoretical value? Set up a sensitivity study that will allow you to create a graph of natural frequency as a function of spring stiffness. (HINT: create a Creo parameter for the stiffness value.)

3.   Can you devise a method to create an assembly which contains a pre-loaded spring?

4.   Call up our 3D centrifuge model created using shells at the end of Lesson 7. Modify this to use cyclic symmetry and compare results with the previous model.

5.   Load our 3D frame model **(frame3d)** from Lesson 8 and find the first four modes of vibration (frequency and mode shape). Plot the mode shapes.

6.   Find out what happens if you try to define a contact surface pair between two surfaces that are in interference. For example, if the pin is slightly larger (by a few 1000^{ths} of an inch) than the hole in our assembly model.

7.   Find the first 4 natural frequencies of a thin (1 mm) circular plate with a diameter of 300 mm. The plate is made of aluminum. How do these frequencies depend on the edge condition: clamped (no rotation) versus free (rotation allowed)? What type of elements must you use to answer this question?

8.   For the two beams model, find the effect of the connecting spring stiffness on the load carried by the upper beam. For the most extreme values of this stiffness (zero and infinite), what load should be carried by the upper beam?

9.   Create a model of a musician's tuning fork. Assume the material is stainless steel. Find a real one and, using its dimensions, see how accurately you can determine its frequency of vibration. (Hint: the computed natural frequencies are very dependent on the constraints.) Is the fundamental mode the desired mode for "tuning"? A common tuning fork used by guitarists produces a tone of the note A above middle C (440 Hz.). How, exactly, does a tuning fork work? How sensitive is its natural frequency to its dimensions? Repeat the analysis using aluminum as the material. What happens to the mode shapes and natural frequencies? (This is not a trick question but has a surprising answer.)

10.   In the contact analysis definition, investigate the options in the **Convergence** tab for *Localized Mesh Refinement* and *Check Contact Force*. Also examine the effect of changing the *Number of Master Steps* under the **Output Steps** option.

This page left blank.

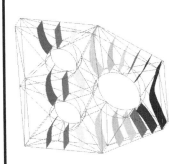

Chapter 10 :

Thermal Models

Steady state and transient models; transferring thermal
results for stress analysis

Overview of this Lesson

This lesson will give you a very brief introduction to the main features involved in
running thermal models in Creo Simulate ("Thermal" for short). We will first have a
quick overview of the program and discuss the different model types available. Then we
will look at the various boundary conditions and heat loads that can be applied to the
model. One of the most confusing aspects of using Thermal is the subject of units - some
material is presented to help organize this important topic.

After these preliminaries are over with, we will look at four examples: a solid model, a
2D plate idealization, a transient (time dependent) problem, and finally the computation
of a temperature field in a solid model to be used to calculate thermally induced stresses.
Most of the underlying procedures for setting up these models, design studies, and result
windows should be familiar to you so they will not be presented in fine detail. There are
some new result window types available for thermal problems, and some opportunities
for making further use of exported result data.

Overview of Thermal Mode

In this section we will present a fairly terse "bird's eye" view of Thermal, including what
problems it can solve, the types of models available, and the variety of boundary
conditions and heat loads that can be applied to the model.

There are two main challenges in thermal modeling. First, you may find that, unlike
stress analysis, your physical intuition is challenged a bit more in anticipating or
interpreting the results. For a lot of users, the realm of heat transfer is not usually familiar
territory. Therefore, you are encouraged to spend additional time when going through the
examples below to explore what happens to the model when you make alterations and to
try to reconcile those results with your physical understanding of the problem. By all
means, make up your own problems based on your own experience.

The second challenge involves the variety of different physical attributes that are either
required or reported. In stress analysis we are interested in load and displacement (stress
and strain), and really only have two physical material parameters (Young's modulus and

Poisson's ratio) to worry about, at least in linear problems. In thermal analysis, although the primary solution variable is temperature, we are often equally interested in heat flux through the model and, in particular, heat transfer through the surface. Unlike stress analysis, where static problems are the norm, heat transfer often involves transient problems which add another layer of complexity. There are numerous parameters that affect these results: material properties (thermal conductivity, density, and heat capacity), surface conditions (specified temperature or insulated), internally generated and/or applied surface heat loads, convection heat transfer at the surface, time dependent loads and boundary conditions, and so on. Keeping track of these factors, and in particular their units, will be a challenge. We will spend some time discussing units a bit later in this introduction.

What can you do with Thermal?

Creo Simulate Thermal is a program that solves the equations involving heat conduction in solids. These solutions can represent a steady state condition, or can be time dependent (transient). The primary solution variable is the temperature of the solid. The solution incorporates or accounts for the physical attributes and processes shown in Figure 1 below. Thermal will solve for the internal and surface temperature wherever it is not explicitly prescribed in a boundary condition.

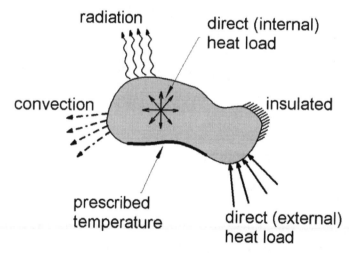

Figure 1 Basic physical attributes and processes in a thermal model

As indicated in Figure 1, the model can be subjected to numerous conditions on its interior and surface:

- **Direct heat addition/removal.** Heat energy can be directly added to (or removed from) the model throughout its interior (think of heat generation in a nuclear fuel rod, or perhaps by microwave heating). Heat can be transferred through its surface by specifying a heat flux which is either uniform or can vary across the surface.
- The surface could be subjected to a **prescribed temperature.** This can be either uniform or vary across the surface.

- In **convective heat transfer**, the temperature of the surface is not known *a priori*, but we specify a convective heat transfer coefficient and the temperature of the surroundings. The surface temperature is then obtained as part of the solution.
- In **radiative heat transfer**, the surface temperature is obtained as part of the solution. Note that radiative heat transfer is a highly non-linear phenomenon.
- **Any model surface not specifically subjected to any of the above is assumed to be insulated.** This means there is no heat flux through the surface (but along the surface is OK), and the surface temperature is computed as part of the solution.

Why use Thermal?

Based on the computed temperature field, Thermal can compute the heat fluxes within the model and through the surface(s). Obtaining these heat fluxes is often the primary motivation for executing the model. For example, it may be desired to design a cooling system for an electronic component. The design goals would be to not exceed a maximum allowed temperature of the component and to obtain a required rate of heat rejection from the device into the surroundings. These goals would be influenced by the geometry, material, and surroundings of the system. Sometimes the temperature distribution itself is of interest, as when it is desired to determine the best location for mounting temperature sensors on a system, and to estimate the relation between the measured temperature at the sensor and the temperature at another critical but inaccessible point. Finally, the temperature of the model can be returned to Creo Simulate Structure as a thermal load which, via thermal expansion, can induce (sometimes very high) stresses in a part or even affect properties like Young's modulus.

Material Properties

The temperature distribution within the solid is governed by Fourier's law of conduction which relates the heat flux through the solid to the gradient of the temperature. Thus, within the solid, the primary material property of interest is the *thermal conductivity*, the key parameter in Fourier's law. All materials in the Creo Simulate materials library have values for this parameter. The default materials are isotropic, although custom materials can be created that have different conductivity in different directions. For transient problems, the ability of the material to absorb or hold thermal energy is governed by another material property, the *specific heat capacity*. This is also stored in the materials library (along with the *mass density*). Finally, when a temperature field is passed back to Structure to compute thermally induced stresses, the solution will depend on the value of the *coefficient of thermal expansion* of the material. Thus, we see that there are four material properties of interest, rather than the two required for Structure.

Model Types and Idealizations

As in Structure, the default model type is a **3D solid**. This is by far the most straightforward model to deal with (see the discussion of Units below!). However, as we have seen, quite often the essence of a problem can be represented by a simplified model type that captures the central features of the problem and can produce useful results much more efficiently (remember the Golden Rule!). There are three additional model types available in Thermal:

2D Plate
- thin flat plate in XY plane
- model with shells created on surfaces
- analogous to plane stress models in Structure

2D Unit Depth
- conduction perpendicular to model negligible
- model with shells and solids
- analogous to plane strain models in Structure

2D Axisymmetric
- on X>0 half of XY plane; symmetry about Y-axis
- model with shells and solids

In this lesson, we will only look at 3D solid and 2D Plate models.

Of course, symmetry can play an important role in the solution of thermal problems, as it did in structural problems. Most often, unless the temperature on a symmetry plane is to be specified, the appropriate symmetry boundary condition is obtained by doing nothing. Since there can be no heat flux through the symmetry plane, it must be treated as insulated, which is the default. Cyclic symmetry is also available in Thermal models.

As indicated above, model idealizations are also possible in Thermal using beams and shells. **Beams** are strictly a 3D idealization shortcut. **Shells** can be used in both 2D and 3D problems. Treatment of these idealizations is quite far beyond the scope of this lesson[1]. There is considerable on-line help available if you need to investigate these further.

More on Boundary Conditions

There are four possible surface boundary conditions that more-or-less directly involve the surface temperature.

Specifed Temperature
The surface temperature is specified. It can be uniform or vary across the surface.

Convection
This condition requires the specification of a convective heat transfer coefficient

---

[1] For example, the definition of a convective heat transfer load on a beam with a specified cross section, but represented as a linear element, is a complex task.

and an external "bulk" temperature. The convection heat transfer can result in heat fluxes that either add or remove energy from the part. The heat flux is related to the surface temperature according to

$$Q = h A (T - T_B)$$

where (in SI units)

Q = heat transfer (W = J/s = N.m/s)
h = convection coefficient (W/m^2.°C)
A = surface area (m^2)
T = surface temperature (°C)
T_B = bulk temperature (°C)

Note that a positive Q results when the surface is hotter than the surroundings.

Radiation

Radiative heat transfer is a non-linear phenomenon. In Thermal, radiation can act only between the model and the surroundings (that is, parts or surfaces in the model cannot transfer energy to another part or surface via radiation). It is governed by the following equation

$$Q = \varepsilon \, \sigma A (T^4 - T_\infty^4)$$

where (in SI units)

Q = heat transfer (W = J/s = N.m/s)
ε = emissivity
σ = Stefan-Boltzman constant (W/m^2.K^4)
A = surface area (m^2)
T = surface temperature (°K)
T_∞ = enclosure temperature (°K)

As for convection, positive Q results from heat energy leaving the surface.

Insulated

Any surface entity that does not have one of the above three conditions applied to it is assumed to be insulated. This provides a boundary condition on the temperature gradient into the model and normal to the boundary - it must be zero. By Fourier's law, there is no heat flux normal to the boundary. The temperature of the boundary is therefore not specified beforehand but is determined as part of the solution in order to satisfy this gradient condition.

More on Heat Loads

Heat energy can be added directly to the model in several ways. These can be either internal or surface loads. Direct heat loads can be applied to volumes, surfaces, or edges. Depending on the model type and entity selected, the interpreted units of the heat load will be different.

A Note about Units

As indicated earlier, the interpretation and proper use of units is one of the more challenging aspects of using Thermal. Of course, if the units aren't used correctly, the results will be meaningless. You must be very clear about units for

- material properties
- convection coefficients
- heat flux
- heat generation

The units for these will change between model types and even entities within models.

---

* * * IMPORTANT * * *

It is strongly recommended that you avoid the use of British gravitational units (IPS) or similar. In thermal problems, this is an invitation to disaster. This system invokes yet another derived quantity (the Btu, aka British Thermal Unit) that serves only to complicate the issue (not to mention the confusion between weight and mass in that system). Try to work only in SI or mmNs unit systems. It is unfortunate that the Creo defaults (unless reset) continue to use the old system of units.

The units shown in Table I apply to 3D models. In particular, notice the difference in meaning between energy, heat transfer (energy per unit time), and heat flux (heat transfer per unit area). The on-line documentation is sometimes pretty loose in using these terms. For example, at various places, the on-line documentation uses the symbol "Q" (instead of "q") to represent heat flux. In the equations above, this is the symbol for heat transfer.

If you solve only 3D solid models, then you can use the units in the table and you probably won't have much trouble. For example, tabulated values of convection coefficients, conductivities, and so on, are reported for the 3D world and correspond to the units shown in the table. These are the values in the material library.

The difficulty in Thermal comes when we set up alternate model types or apply loads in different ways. For example, when we apply a convective heat transfer coefficient to the edge of a 2D Plate model, in order to account for the thickness of the plate model, the coefficient must be expressed (in mmNs system) in units of mW/mm.°C. Similarly, an interior heat load in a 3D model applied to a beam element must be expressed in terms of power per unit length of the beam (mW/mm). To create these alternate unit versions of coefficients and loads, we must alter their numeric value. If this is not done correctly, of course our results will be wrong. We will explore this aspect of Thermal in the first two models.

For more information on the use of units, see the extensive on-line help. This also contains information on unit conversion between systems.

Table I. SI units in Creo Simulate Thermal (for 3D models)

Quantity	Units (mm - N - s)	Units (m - kg - s)
length	mm	m
mass	tonne	kg
density	tonne/mm^3	kg/m^3
force	tonne.mm/s^2 = N	kg.m/s^2 = N
energy	N.mm = mJ	N.m = J
power, heat transfer (Q)	N.mm/s = mJ/s = mW	N.m/s = W
heat flux (q)	mW/mm^2	W/m^2
thermal conductivity (k)	mW/mm.°C	W/m.°C
convection coefficient	mW/mm^2.°C	W/m^2.°C

Well, that's probably more than enough preliminary discussion. Let's get started on our first model.

Steady State Models

3D Solid Model

Launch Creo and create a new folder [**chap10**]. Create the model called [*solid_plate*] shown in Figures 2 and 3. Make sure you use the *mmNs* system of units. Sketch the section on FRONT and extrude backwards so that the front face is on the XY plane of the default coordinate system (we will need this geometry for the plate model later). The plate is 50 mm thick.

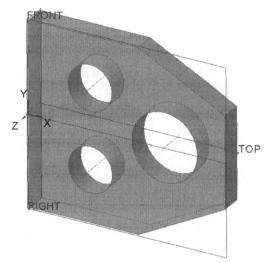

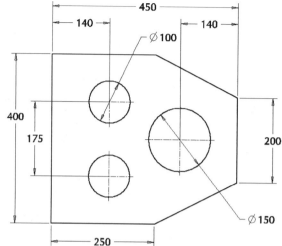

Figure 2 Solid model (note location of default CSYS)

Figure 3 Section shape for model #1 (units mm)

Save the model and transfer into Thermal using

Applications > Simulate > Thermal Mode

Open the **Model Setup** button to see the options there. We will not model any assemblies in this lesson but should note in passing what interfaces might be available.

You will notice a few new icons appear in the ribbon replacing some of the tools used in Structure. These are shown in Figure 4. In the **Boundary Conditions** pull-down menu the **Symmetry** tool allows cyclic symmetry[2]. Other tools in the Home ribbon are the same (**Materials**, **Measures**, etc.).

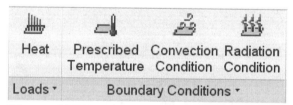

Figure 4 THERMAL toolbar buttons

Assign the material copper (CU) to the part. Note the orientation of the WCS. You can turn off all the datums since they will not be needed.

Select the **Prescribed Temperature** button. The boundary conditions in Thermal are also organized in sets. The first one is BndryCondSet1. Enter a name like *[t500]* for the prescribed temperature. The **Surfaces** reference type is already selected. On the model, pick the surface on the left face of the plate. In the **Value** field, enter a value **500** (°C). Open the **Advanced** command and check out the other available Spatial Variations. There are options here to let you specify a prescribed temperature that is a function of position, or interpolated over the chosen entity. Leave this set to **Uniform**. See Figure 5. Accept

---

[2] As mentioned previously, a planar symmetry condition is applied by doing nothing (i.e. surface behaves as if insulated).

the dialog window and a prescribed temperature icon will appear on the model.

At the narrow end of the plate, we will place a convection boundary condition. Select the **Convection Condition** button. Enter a name for the condition; note that it is still in the same set as the prescribed temperature. Pick the right end surface of the part. Leave the spatial variation set to **Uniform**, and enter a value for the convection coefficient h = **5**. Confirm that the units shown are the standard (mW/mm^2.°C). Leave the bulk temperature set to zero - this is the temperature of the medium receiving the heat energy from this surface (see the equation above). Notice that we could set a time varying bulk temperature if desired. The completed dialog is shown in Figure 6. Accept the dialog. A new convection icon will appear on the right face of the model.

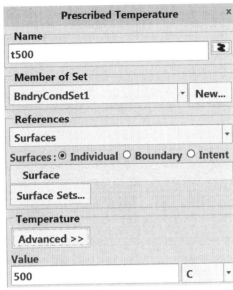

Figure 5 Creating a prescribed temperature boundary condition

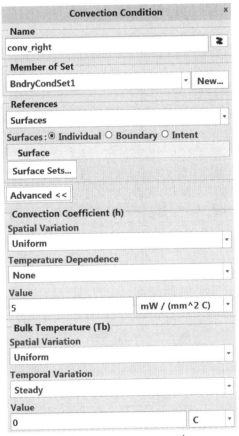

Figure 6 Creating a convection condition on a surface

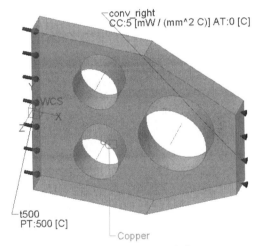

Figure 7 Final 3D plate model

All the other surfaces of the part that we have not set conditions on are assumed to be insulated. The completed model is shown in Figure 7.

Now, create the study with

Analyses and Studies
File > New Steady State Thermal

Enter a study name *[plate3D]*. The constraint set is already selected. There is no thermal load set since there are no applied heat fluxes or heat generation entities. As usual, set up a *QuickCheck* analysis. Change the Output plotting grid to 10 (this will give us smoother contours). Verify your *Run Settings* and *Start* the run. Look in the **Summary** listing for errors and/or warnings. AutoGEM will create something like 80 elements. Near the bottom, notice the new measures that are reported (heat flux, temperature gradient, max and min model temperatures). Assuming no errors, *Edit* the study to run a Multi-Pass Adaptive run with a convergence of 5% (what is it converging on?) and a maximum edge order of 6. The MPA run will converge on pass 4 with a maximum edge order of 5. The minimum temperature is about 22°. The maximum is, of course, 500°.

Now to create some result windows. Open the *Review Results* window. Insert a window definition to create a fringe plot of the temperature - this is the default. Under the *Display Options* tab, select *Show Element Edges*. Select *OK and Show*. See Figure 8. Notice that the fringe legend color palette is a bit different from Structure. Select *Model Min* to locate and label the minimum model temperature on the right end.

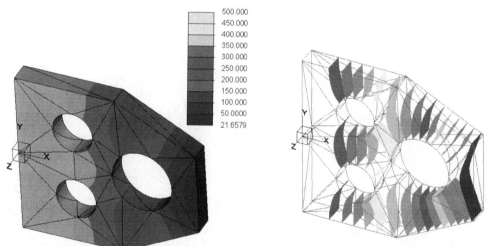

Figure 8 Temperature fringe plot

Figure 9 Temperature isotherm surfaces

Copy the temperature fringe plot. Name the new window *[temp_isosurf]*. Go to the **Display Options** tab and select *Contour*, and *Isosurfaces* and set *Legend Levels* to **15**. Show this window (Figure 9). Observe that the isotherms are basically straight through the plate, as anticipated (all cross sections through the thickness should be identical). If you zoom in, you may note a regular texture on some of the isosurfaces. This is clearly a numerical artifact and is not real. You can use

Format > Edit

to change the color spectrum for the isosurfaces.

Now create a new window called *[flux]*. Select the following:

Display Type (Vectors)
Quantity (Flux) | Component (Magnitude)

In the **Display Options** tab, select the *Animate* option. Show the window (Figure 10). What you see now is an animated representation of the heat flux through the part. Notice that the flux vectors are scaled and colored according to the magnitude of the local heat flux. Also, the vector direction is perpendicular to the isotherms (a consequence of Fourier's law).

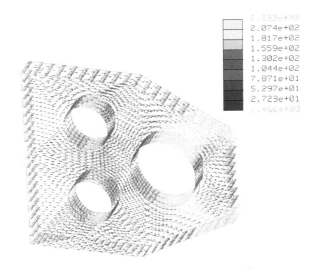

Figure 10 Heat flux vectors

Go to the FRONT view of the flux vector window. If you zoom in on any of the insulated edges, you should expect to see the flux vectors running tangent to the surface. Any indication of a flux vector not tangent to the surface represents a numerical or modeling error. We will examine this in more detail in the next 2D model.

For completeness, create result windows showing convergence plots of the following:
  ▸ maximum temperature (constant in this case)
  ▸ minimum temperature
  ▸ maximum heat flux

Have a look at these, then use *File > Save as > Save a Backup* to create and save a result window template called *[MPA_3DThermal]*. This will come in handy later.

We have now seen the main features of steady state thermal modeling for a 3D solid model - boundary conditions, convective loads, and prescribed temperatures, and some new variations available in the result window. We will now have a quick look at a different model type, and some additional result review tools.

2D Plate Model

Close all the result windows and return to the main display. Save a copy of the current part file as *[solid_plate_3D]*. Notice that this does not change the name of the part file currently in session which we now convert to a 2D model. In the **Home** ribbon select

Model Setup > Advanced

Select the **2D Plane Stress (Thin Plate)** option. Pick on the default coordinate system, then pick on the front surface of the part (note this is in the XY plane). Accept the dialog. Because we are changing model type, we must confirm the change (since a number of modeling entities are lost). The surface is now highlighted in purple.

We must recreate the boundary conditions on the prescribed temperature and convection ends. Before we do that, we need to create a shell on the chosen surface. This is identical to the creation of a model for plane stress analysis. In the **Refine Model** ribbon, select the **Shell** button, then click on the front surface of the part. In the **Shell Definition** window, set a thickness of 20 (mm). The material may already be defined, otherwise bring in the material copper. Accept the dialog. The shell idealization will be listed in the model tree. Notice that, as for the plane stress models we dealt with earlier, the 2D shell does not pick up its thickness from the 3D model but is specified independently (recall that the thickness of our Creo model is 50mm). Open the **Simulation Display** window and see the effect of the **Shells** check box on the **Modeling Entities** tab.

Select the **Prescribed Temperature** button. Call it *[t500]* as before. Set the **References** type to **Edges/Curves**, pick on the front left edge of the plate[3]. Enter the value **500** and accept the dialog.

The creation of the convection condition is essentially the same, with one unexpected twist due to the way Thermal handles convection coefficients on edges. Select the **Convection Condition** button. Enter a name, like *[rightedge]*. Under **References**, select the **Edges/Curves** type and pick on the right edge of the shell. Click the **Advanced** button. See Figure 11. Leave the spatial variation set to **Uniform**. For the value of the convection coefficient, we must be very careful.

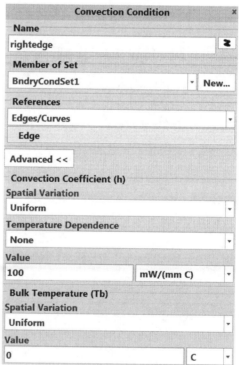

Figure 11 Setting a convection coefficient on a curve in a 2D model

Recall that for the previous 3D model, our value of the convection coefficient was 5 mW/mm^2.°C. This represents heat flux (mW) per degree per unit area of the surface. In the 2D plate model, we are specifying a coefficient along a linear entity only, and our units will be heat flux per degree per unit length along the edge. We must therefore account for the thickness of the plate - each millimeter along the edge corresponds to 20 mm^2 of actual plate surface. Thus, we must **multiply the 3D heat flux value by the plate thickness**. So, in the dialog window, enter a value of 100 for the convection coefficient (note the units). Leave the bulk temperature set to 0.

---

[3] If the shell idealization is being displayed as a shaded solid, note that you cannot select any of its shaded surfaces or edges.

Our model is now complete (see Figure 12). What symbol seems to be missing from this view? Is that a problem? The missing symbol is the material assignment, but remember that this is part of the shell definition.

Go ahead and create a new steady state thermal design study *[plate2D]*. Set up a **QuickCheck** analysis with a plotting grid of 10. Run the analysis and look for messages in the **Summary** list. How many elements does AutoGEM create? If all looks satisfactory, **Edit** the analysis and change it to an MPA with 1% convergence, max order 6. Run the study.

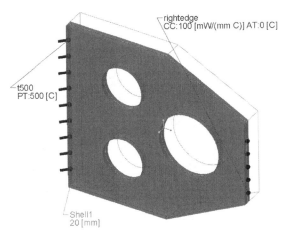

Figure 12 Completed 2D Plate model on front surface of part

The MPA converges on pass 6 with a maximum edge order of 6. The minimum temperature reported is about 21 °C - about the same as the solid model solution.

Create some result windows to show the temperature fringe plot and the flux vectors. Although this is a 2D model, the previous template (**MPA_3DThermal.rtw**) will work just fine.

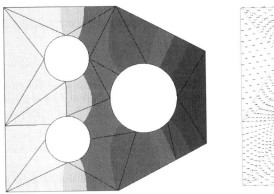 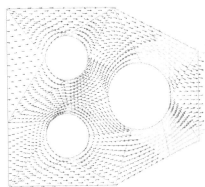

Figure 13 Results from 2D Plate model - temperature fringe (left) and flux magnitude vectors (right)

We should have a look at some data to study the behavior of the solution. Create a couple of convergence graphs - for example for the maximum flux magnitude and the minimum model temperature. These are both built-in **Measures**. These will look something like Figure 14.

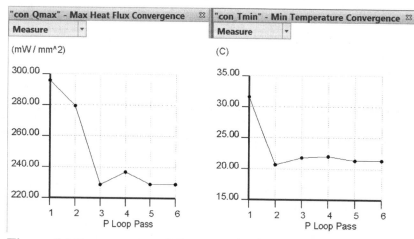

Figure 14 Convergence of flux magnitude and minimum temperature in 2D Plate model

Let's have a detailed look at the temperature along the right edge. Set up another graph as follows:

> *Display Type (Graph)*
> *Quantity (Temperature)*
> *Graph Location (Curve)*

Select the button under *Curve*. A new window shows the model (Figure 15). Click on the right vertical edge. Middle click and note the message that indicates how the graph will be set up. Accept this, then show the result window. See Figure 16. The temperature minimum is at the center of the plate, and the variation is essentially symmetrical as expected. According to this temperature distribution, what direction should the heat flux component in the Y direction be going along this edge? Create a graph window that shows the Y component of the flux along this edge to confirm this. (Hint: you can easily change the window contents by using the drop-down lists at the top left corner of the result window.)

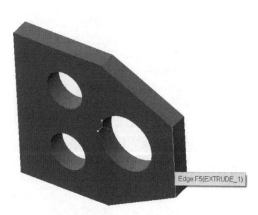

Figure 15 Selecting a curve for the temperature graph

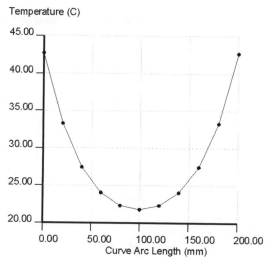

Figure 16 Temperature along right vertical edge

We will now set up another graph to perform an auxiliary calculation on this solution data. We would like to find the value of the net heat transfer coming out the right end of the model - this is the heat transferred to the surroundings. We can obtain that from the heat flux leaving the right edge of the part. We are interested only in the X component of the flux, since the Y component runs parallel to the surface.

Copy the temperature graph result window definition to another graph called *[Xflux]*. Change the **Quantity** to *Flux*, and **Component** to *X*. Select the same curve we had previously and show the graph. This graph is also reasonably symmetrical as expected with an average value of about 135 mW/mm^2. With the graph display window highlighted, select

File > Save As > Save a Backup

and check the Type drop-down list in the lower right corner. This gives a number of different formats for the exported data. If you export it in spreadsheet form (MS Excel xls), you can easily modify the spreadsheet to integrate (perhaps by using trapezoidal integration) the flux with respect to the length along the edge. This value represents the total heat transfer per unit thickness of the plate[4]. Once again you must be very careful with units. The flux data is given in units of (mW/mm^2). The integral of flux with respect to edge length then has units of (mW/mm). Multiplying by the thickness of the plate finally gives us the heat transfer out of the plate. The value should be approximately[5] 540 Watts, about a third of a toaster!

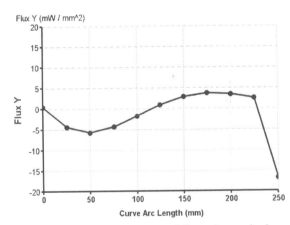

Another interesting graph to plot is the Y component of heat flux coming out the top horizontal edge of the model. Since this edge is insulated, there should be no heat flux through the surface (ie in the Y direction). Create this graph. See Figure 17. Observe that the flux is not, in fact, exactly zero which represents a small modeling error. It is considerably smaller than the X component, however.

Figure 17 Y component of flux through the top edge (should be zero)

You might like to find out how to plot the flux in a radial direction through the edges of the circular holes. This should also be zero.

Evidence of flux components normal to insulated surfaces is reasonable grounds to try

---

[4] There is actually a built-in **Measure** for this calculation. See **Heat Transfer Rate**.

[5] The total heat transfer is approximately (average flux value) * (edge length) * (plate thickness) = 135 mW/mm^2 * 200mm * 20 mm = 540 W.

some mesh refinement. In the above, we have used all AutoGEM defaults. It is left as an exercise to revisit some of these models to see what the effects of changing the AutoGEM limits, or adding some mesh controls, would be. As always, we are interested in finding out how the model responds to various settings and run options. Also, noting that both the 3D and 2D models are symmetric, you should cut the model in half in Creo Parametric and rerun the studies. Do you really have to do nothing on the symmetry plane? What about the other boundary conditions?

Transient Analysis

A transient, or time dependent, analysis will allow you to find out how quickly (or slowly) the model reacts to the given boundary conditions. If the boundary conditions are constant, the transient analysis will eventually arrive at the steady state solution (as we calculated in the previous models). If the boundary conditions are also time dependent, then a transient analysis lets you observe the complicated time-dependent response of your model.

There are some restrictions on what you can do with a transient analysis. First, only 3D solid models can be treated. No shell or beam elements are allowed. The material(s) must be isotropic.

We will do a simple example of a transient analysis. The model consists of a long thin steel rod (300 mm long, 30 mm diameter). This model is shown in Figure 18. Create that now, making sure your units are set to mmNs and the rod is along the x-axis.

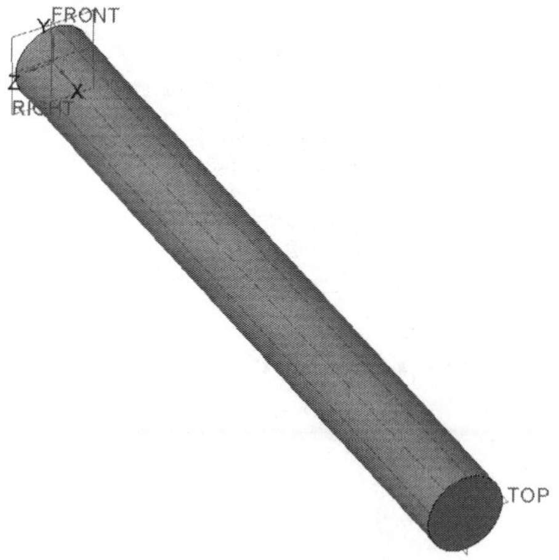

The rod is initially at a uniform temperature of 20°C. One end of the rod (x = 0) is maintained at the initial temperature. Starting at time t = 0 seconds, the other end of the rod (x = 300mm) is exposed to a very hot (1000°C) environment, through a convection heat transfer coefficient of 0.5 mW/mm². °C. The remaining cylindrical surface of the part is assumed to be insulated. The primary questions we want answered

Figure 18 Circular steel rod for transient analysis

concern the eventual temperature of the hot end of the rod, and the time it takes to reach that steady state.

Once the part is made, transfer into Thermal with

> ***Applications > Simulate > Thermal Mode***

Assign the material **STEEL** to the part.

Define a constant temperature surface at the left end (x = 0) using the ***Prescribed Temperature*** tool. Call the condition *[tcold]* in BndryCondSet1. The default **Reference** is **Surfaces**. Pick on the left end of the rod, and enter a value of *20* (°C).

Define a ***Convection Condition*** on the right end (x = 300mm). Call it *[hotend]* in BndryCondSet1. Pick the surface at the right end. Enter a convection coefficient of *0.5* (note the units) and a bulk temperature of *1000 C*. See Figure 19.

The model is now complete for a steady-state analysis. Open the ***Analyses and Studies*** window and select

File > New Steady State Thermal

Call the analysis *[trans_rod_SS]* (for steady state). Set up a ***QuickCheck*** analysis, with a plotting grid of *10*. Check the run settings, and ***Start*** the study. Open the **Summary** window and look for errors. If all is satisfactory, ***Edit*** the study to run an MPA with a convergence of 1%, max order 6. Run the new study. The run converges on pass 3, max edge order 3. The maximum temperature in the model is reported as 782°.

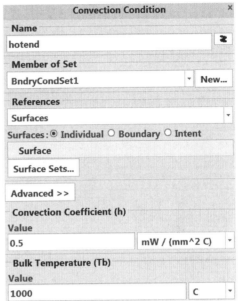

Figure 19 Defining the convection condition

Create and show a result window for the temperature fringe. See Figure 20. This shows a series of equally spaced fringes. Recall that the default fringe legend uses equally spaced values. This fringe pattern indicates a linear temperature variation along the rod (Does this make sense?). This is confirmed by creating another result window that shows a graph of temperature along a curve extending the length of the centerline of the rod (Figure 21). Come back later to check this out.

Figure 20 Steady state temperature fringe plot

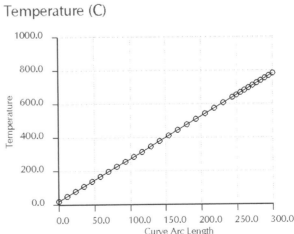

Figure 21 Variation of temperature along rod at steady state

Now for the transient analysis. Close down all the result windows and return to the *Analyses and Studies* dialog. Select

File > New Transient Thermal

Call this analysis *[trans_rod1]*. Enter the description (see Figure 22). Select the constraint set. In the **Temperature** tab, set the initial temperature value to *20*. Notice that the **Distribution** option allows you to specify a previously calculated temperature distribution for the initial condition (MecT refers to MECHANICA Thermal, the previous name for Creo Simulate Thermal) .

Open the **Output** tab. We will leave all defaults here. This is to use automatic time steps from the start of the run (t = 0) until the steady state condition is reached (note the check beside **Auto**). Notice on the Convergence tab the only options are **QuickCheck** and **Single-Pass Adaptive** (or **SPA**). Accept the analysis definition.

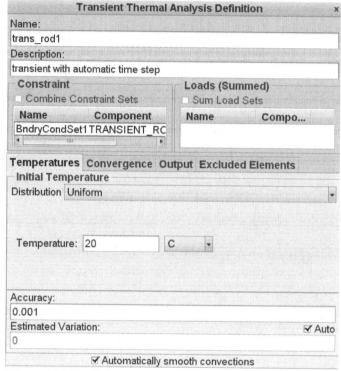

Figure 22 Defining a transient analysis

Start the transient analysis and open the **Summary** window. The solution proceeds through a number of (unequal size) time steps. If you scroll to the top of the window, you will observe the edge order used for the solution, and the size of the first few time steps (quite small!). As the solution proceeds, the time step is allowed to increase. Observe that every few time steps, the maximum edge order is adjusted (sometimes downward!). This is to account for the changing temperature distribution within the rod. The last time step starts at about 12,000 seconds and the reported final temperature variation is 762C. This means the temperature on the hot end is (20C + 762C =) 782C, the same as computed using the steady state analysis done previously. Check this by creating a measure for the maximum temperature on the end surface[6] and re-run the study.

---

[6] In the **Measure Definition** window, check the *Time Eval* box to access the data for a transient analysis.

Create a result window. The default (and only) option is to show a graph. Select the button under *Measure*, and select **max_dyn_temperature** in the list of defined measures. Show this graph - see Figure 23. This shows that the solution changes very rapidly for the first few seconds, but by 5000 seconds has essentially reached its final value. Leave the result windows.

If you mouse-over one of the data points on the graph a pop-up will give you the data for that point. Suppose you want to find the temperature at a precise time. One way to do that, at least partially, is as follows.

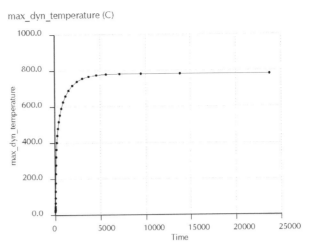

Figure 23 Maximum temperature in transient solution

Create a new transient thermal analysis called *[trans_rod2]*. Set the **Temperature** tab as before. In the **Output** tab, in the **Output Steps** list, select *User-defined Output Steps*. See Figure 24. Set the number of steps to *41*. Then select *User Defined Steps*. Scroll down to the bottom of the list of points, enter a time value of *10000*, then press *Space Equally*. This will yield equal time steps of 250 seconds.

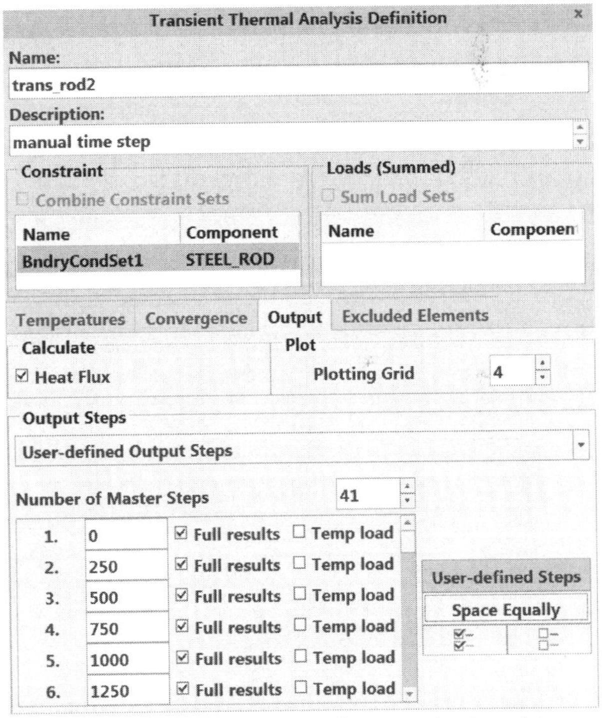

Run the new analysis and open the **Summary** window. You will observe that the number of actual steps is larger than we requested. However, on closer examination, notice that the run did stop at 10,000 seconds, with most of the time steps set to 250 seconds as requested. Early in the run, however, the time steps were considerably smaller in order to track the rapidly changing solution. The actual computation time steps are controlled by Thermal using a different setting in order to maintain accuracy. Very nice to know!

Figure 24 Creating a transient analysis with equal spaced intervals

Create a result window to show a graph of the maximum dynamic temperature (a built-in **Measure**). See Figure 25. In the result window definition dialog, notice that you can select any particular time step and plot data for that particular instant in time. Observe that in Figure 24, the boxes "Full Results" were automatically checked for each time interval. Another interesting output result is obtained if you request an animated fringe of the temperature. Select ***Contour*** and ***Isosurfaces***. When you show this, the isotherms will move down the rod, eventually reaching their equally-spaced steady state positions. The final configuration is shown in Figure 26.

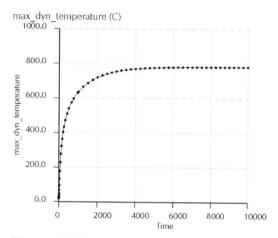

Figure 25 Maximum temperature in transient solution (equal spaced intervals)

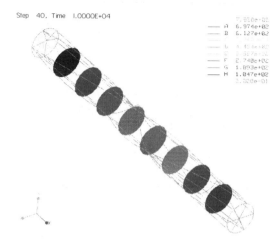

Figure 26 Isotherms at end of transient solution

You might also have a look at the flux magnitude in the rod at various time steps. Initially, the magnitude is essentially zero everywhere. At the final time step, the magnitude is essentially constant everywhere.

We will leave transient analysis now. One thing you might like to try on your own is to perform a similar transient analysis on the rod with a specified heat load on one end (replacing the convection) where the heat load is a function of time. How is the time-dependent response affected by the material heat capacity and thermal conductivity? These are the kind of effects that make transient heat transfer so complicated (and interesting!).

Save the model and remove it from your session.

Thermally Induced Stresses

Our final example in this chapter is to show how a computed temperature distribution can be used to calculate thermally induced stresses. The temperature field affects the model through the effect of thermal expansion. If the model is physically constrained against this expansion (or contraction), very high stresses can be produced. If the model is not constrained but the temperature varies over the model, this can also induce very high thermal stresses.

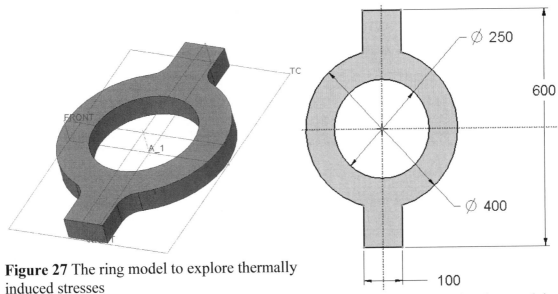

Figure 27 The ring model to explore thermally induced stresses

Figure 28 Section shape for ring model (units mm). Thickness = 50 mm

We will examine the simple model shown in Figure 27. Create this model called *[ring]* using the mmNs part template. The sketch for the shape is shown in Figure 28. Don't forget to add the four rounded corners (R50).

Creating the Thermal Model

When the model is created, transfer into Thermal using

Applications > Simulate > Thermal Model

We will load the thermal model with two prescribed temperature surfaces and a convection condition on the inner cylinder.

First, select the *Prescribed Temperature* tool. Name the condition *[hotend]* and pick on the farthest front rectangular surface. Assign a temperature of *100* (°C).

Select the *Prescribed Temperature* tool again and create a condition *[coldend]*. Pick on the farthest back rectangular surface and assign a temperature of *0*.

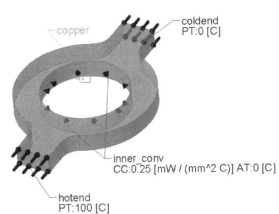

Figure 29 Thermal model ready for analysis

Select the *Convection Condition* tool. Name this condition *[inner_conv]*. Pick the inner cylindrical surface. Apply a convection coefficient of **0.25** (what are these units?) with a bulk temperature of *0*.

Finally, assign the material CU (copper) to the part.

The model is now finished (Figure 29) and ready for analysis.

Create a *New Steady State Thermal* analysis called *[ring_temp]*. Set up a QuickCheck analysis. This should create a mesh of about 70 elements.

---

* * * IMPORTANT * * *

The same mesh must be used in both Thermal and Structure (in order to align the temperature data with the structural elements). For subsequent runs of this model, make sure the **Run Settings** option *Use Elements from an Existing Study* is checked. If you change the model geometry, however, you will have to check *Create Elements during Run* the first time you re-run any analysis in either Thermal or Structure mode.

Assuming the QuickCheck succeeded, change to an MPA analysis with a convergence of 5%. Check the *Run Settings* and *Start* the analysis. The run should converge on pass 3 with a maximum edge order of 4. Nothing unusual here.

Create a temperature fringe plot. See Figure 30. The hot end shows as 100° with the cold end at 0° as desired. Notice that this model has two planes of symmetry. You might like to come back later and implement symmetry conditions (insulated surfaces) on a quarter model[7]. Create a vector plot showing the flux magnitude and direction. Does this data make sense (check the surfaces around the central hole)? Close the result window.

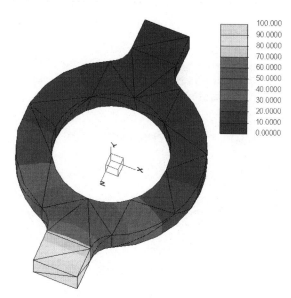

Figure 30 Temperature distribution on ring model

We are now ready to transfer to Structure where we will apply this temperature distribution as a thermal load.

---

[7] Of course, if you plan on using the temperature distribution in a Structure model, all aspects of that must also be symmetric.

Creating the Structure Model

In the **Home** ribbon, select

Structure Mode

First we will apply some displacement constraints. Select the ***Displacement Constraint*** tool and create a surface constraint *[cold_end_fixed]* on the cold end. Fix all degrees of freedom. This prevents rigid body motion of the ring, so it is sufficiently constrained.

To see the effect of thermal expansion, select the ***Displacement Constraint*** tool again and pick the hot end rectangular surface. We will constrain this against translation normal to the surface (Z direction) but will FREE the X and Y translations. We are going to set the reference temperature for the model at 0°C, so at the hot end the rectangular section will want to expand in both X and Y directions. If we constrained against this motion, we would find extremely high stresses induced here (try it later!) that would overshadow what is happening in the ring itself.

To apply the thermal load, we must use the **Loads** group pull-down menu[8]:

Loads > MEC/T Load

This opens a new dialog window. Name the load *[temperature_load]* in LoadSet1. The box beside ***Previous Design Study*** is checked. The study *ring_temp* that we did in Thermal has already been identified for us. The ***Reference Temperature*** at the bottom indicates the temperature at which there is zero stress induced in the model consistent with the current constraints. Leave that set to 0. The completed dialog window is shown in Figure 31.

Open the model tree, and select

Settings > Tree Filters

Pick the **Simulate** tab and turn on the check beside **Thermal entities in Structure** then *Apply*. You can now see all of the Simulate-related data listed there. Now open the **Simulation Display** window and in the **Loads/Constraints** tab turn on all the checkboxes under **Thermal**. The model now shows all the boundary constraints, boundary conditions,

Figure 31 MEC/T dialog to assign a thermal load from a previous analysis

---

[8] The **Temperature Load** button in the ribbon is for global temperature loads only, that is, a constant temperature everywhere.

and loads as in Figure 32. Although you can get information about the thermal data at this time, you cannot edit or delete them.

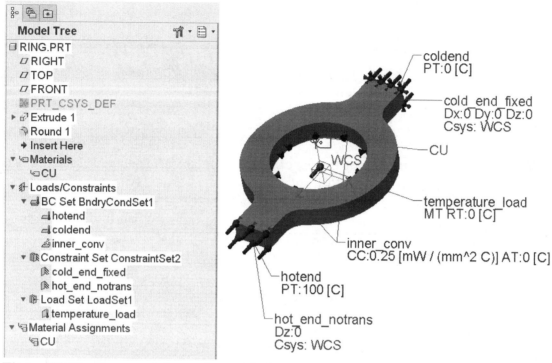

Figure 32 Structure model complete with thermal load

Create a **New Static** analysis called **[ring_stress]**. Make sure the constraint and load sets are selected. Set up a **QuickCheck**. Make sure the **Run Settings** are correct (recall the note above about using the same mesh as the thermal analysis) and **Start** the run. Assuming there are no errors, change the analysis to an MPA with a 5% convergence and rerun the study.

Open the **Summary** window. The analysis just converges on pass 5 with maximum edge order 6. The reported maximum Von Mises stress is about 110 MPa. The maximum displacement in the X direction is about 0.2 mm. This is not much on something 400mm in diameter - it doesn't take much thermal load to produce fairly high stresses.

Create a couple of result windows that show the Von Mises stress fringe plot, and the animated deformed model. These are shown in Figures 33 and 34 below. Locate the position of the maximum stress in the model. If you view the model from the top, an animated deformation will maybe give a hint as to why the maximum stress occurs where it does. What do you think would happen if we decreased the size of the rounds (or deleted them altogether)? Observe the expansion of the rectangular hot face. What would happen if we constrained this face against this expansion?

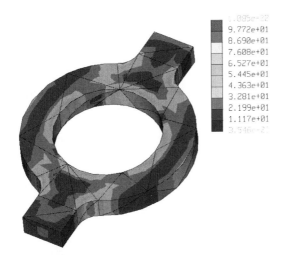

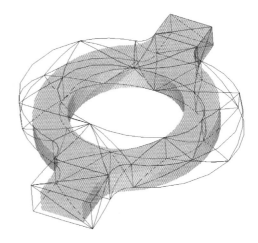

Figure 33 Von Mises stress due to thermal load

Figure 34 Wireframe mesh showing deformation due to thermal load. Note expansion of hot face.

Summary

This concludes our discussion of Thermal mode. This lesson has shown you the basic procedures and tools available for performing steady state and transient thermal analysis. We have explored only two (3D Solid and 2D Plate models) of the four available model types, having missed the 2D Unit Depth and 2D Axisymmetric models. The unit depth model is analogous to the plane strain models in stress analysis, so many ideas should carry over from there. Axisymmetric models have similar restrictions as in stress analysis.

We looked mostly at steady state models here. Transient analysis can become considerably more complicated with the application of time dependent loads and boundary conditions. As usual, there are quite a few tools available that we have not discussed. Hopefully, having seen the general procedures presented here, you will be able to make sense of those tools as you explore them on your own.

As mentioned earlier in the lesson, developing a physical intuition for thermal analysis is a challenge since many people do not have a broad experience in this field. Furthermore, it is necessary to be very careful about units and data entry. It is quite easy to create program input without proper units and not be able to recognize this error in the program results. A lot of care and experience is necessary here.

Conclusion

This completes this tutorial, and we have covered a lot of ground. Even so, we have not looked at many Creo Simulate commands, functions, or analysis types. It is hoped, however, that you are now comfortable enough using the program that you can experiment with these other capabilities on your own without getting lost and that you can more easily follow the on-line documentation. When you do try something new, you should set up a simple problem for which you already know the answer (either quantitatively or qualitatively) just to make sure your procedures are correct. Also, the on-line documentation is available to answer your questions about other aspects of Simulate.

Your installation may also have a set of verification examples installed with the software. See the online **Help** for discussion of the **Verification Guide**. These problems are fairly simple and cover almost all the functionality of Creo Simulate, including the following kinds of models:

Linear
- static analysis
- modal analysis
- thermal (steady state and transient) analysis
- dynamic time and frequency analysis
- buckling analysis

Non-linear
- 2D and 3D contact analysis
- prestress modal analysis
- large deformation analysis

The part files and models can be found in the directory

<load_point>/Common Files/mech/ms_verf

Results of these analyses can be compared (some results are in the **Verification Guide**) to analytical solutions where possible and other FEA programs where not. Before you tackle a new problem in any of these areas, you might have a go at one of the verification examples to make sure you are setting up the model properly.

When using Creo to create solid models for simulations, remember the discussion of Lesson 1. As we have seen many times here, the FEA model is not necessarily identical to the CAD model, in fact it is usually not even close! You may be able to defeature the part[9], and certainly use symmetry whenever you can, in order to produce an efficient as

---

[9] Defeaturing does not necessarily mean loss of data. Modern design practice, in fact, involves the use of FEA much earlier in the design process in order to do preliminary concept evaluation. In that case, it is likely that low level or finer detail has not been determined or included in the model yet.

well as effective FEA model. You should also be able to use idealizations to model some (or all) aspects of the problem at hand. These can greatly speed up the computation.

In closing, it is useful to remind ourselves of some of the comments made in the first two chapters. First,

> *"Don't confuse convenience with intelligence."*

Creo Simulate is undoubtedly a very powerful analysis tool but it will not automatically solve your problems for you. You should realize by now that, like all other computer tools, unless it is used properly the results it produces can be suspect. Remember that in FEA we are finding an *"approximate solution to an idealized mathematical model of a simplified physical problem."* It is expecting a lot to hope for results that *exactly* match the solution found by nature! The most we should hope for are answers that are sufficiently accurate so as to be useful.

Second, Creo Simulate is a huge program that will take many, many hours to master. As your knowledge and experience grows, applying your new skills in increasingly more complicated problems is an inviting prospect. However, when you start to feel the urge to rush off to your computer to tackle a new problem, remember the Golden Rule of FEA is to

> *"Use the simplest model possible that will yield sufficiently reliable results of interest at the lowest computational cost."*

It may even be that your problem can be solved in other (cheaper and quicker) ways.

In short,

> *"Let FEA extend your design capability, not define it."*

Good luck, and happy computing!

Questions for Review

1.   What is Fourier's Law? What material property does it involve? What are its units?
2.   What is the (mathematical) consequence on the computed temperature distribution in a solid of specifying the following boundary conditions:
        (a) an insulated surface
        (b) a surface with a specified heat flux
        (c) a surface with convection
3.   What four material properties are involved in a thermal analysis? Make a table of these and show their units in mmNs and SI.
4.   What are the restrictions imposed on performing a transient analysis?
5.   How do you ensure the same mesh is used for both Thermal and Structure modes? Does it matter which mode creates the mesh?
6.   What interface types are available for thermal modeling of assemblies?

Exercises

1.   Explore the effects of various mesh control functions on the 3D Solid and 2D Plate models. In these models, describe the effect of mesh control on execution time, convergence, surface fluxes, max/min temperature, and so on.

_____

2.   Cut the third 3D model in half along the horizontal symmetry plane. What boundary condition is required to implement symmetry for the Thermal and Structure models? Do the results agree with the full model?

_____

3.   Find out if thermal problems are subject to the same sorts of singularity problems that occur in stress analysis. For example, if two plane surfaces meet along an edge, what happens to the flux at the corner if you specify different temperatures on the two surfaces?

_____

4.   How could you set up a model that would illustrate the effect of the material's specific heat capacity in a transient analysis? Try it! Does this property have any effect in a steady state analysis?

_____

5.   Create a model of a solid sphere. The sphere surface is subjected to convection. Perform a transient analysis from an initial uniform temperature state to a final state where the surface temperature is exactly half the initial value. How much of this solution can be verified by analytical means?

_____

6.   Find a built-in Measure that can report the energy transfer in or out of a surface, or the entire model. Try this out on the 3D and 2D plate models. Does the energy

coming out match the energy going in? Should it?

---

7. Redo the transient analysis of the long thin rod but with a convective heat transfer
at the cold end. Use a bulk temperature of $0°$ and a convective heat transfer
coefficient of 1.0 mW/mm^2.$°$C. Plot a graph of the surface temperature on the cold
end vs time.

---

8. This is quite a bit more complicated and will require you to use a **2D Unit Depth**
model. Be very careful of your units!

An in-floor heating system consists of a loop of
plastic tube buried just under the surface of the
concrete floor. The overall geometry is shown
in the figure at the right; details are shown
below (dimensions in millimeters). Neglect all
end effects.

Hot glycol at $60°$C is pumped through the tube
where the estimated convection coefficient is 80
W/m^2.$°$C to the inside surface of the tube. The
tube has an inner diameter of 15mm and an outer diameter of 20mm. The tube material
has a thermal conductivity of 0.4 W/m.$°$C. Assume the bottom of the concrete slab is at a
steady ground temperature of $5°$C. The concrete has a thermal conductivity of 1.4
W/m.$°$C. The top surface of the concrete slab is exposed to the air inside the house at
$20°$C with a convection coefficient of 5 W/m^2.$°$C.

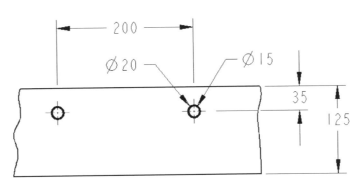

(**Hint**: Seriously consider the use of symmetry here. As mentioned, set this up as a **2D
Unit Depth** model of a single tube. You can use datum curves to divide the model
surface into regions. Apply different material properties to each region as required.)

Consider the following scenarios:

i) Plot the temperature distribution on the concrete upper (floor) surface. Compare this
to generally accepted standards for comfort level (Is there such a thing?).

ii) Find the maximum temperature in the tube and the total heat transfer through the tube wall **per meter** of tube length. How much heat escapes out the bottom of the concrete slab into the ground? How much actually goes into the room? Is all the energy accounted for? What is the efficiency of the system (efficiency = heat into room / heat provided by tube)?

iii) How is the efficiency affected by the horizontal spacing between tubes, and the depth of the tube in the concrete slab?

iv) Is modeling the plastic tube critical to this model or could that be left out to simplify the model? Would your answer change if the tube was made of copper (much higher conductivity)?

v) Find out how much the efficiency of the system will improve if a 40mm thick layer of styrofoam insulation (k = 0.015 W/m.°C) is added below the concrete slab. The bottom of the styrofoam will now be at the ground temperature (5°C). What is the new maximum floor temperature?

Do you think any of these models adequately simulate the actual concrete slab floor? Has this model provided *any* useful information? What effects have not been included? How would you extend this model (what would you add, and what additional information could you obtain)?

---

9. Another more complicated problem, this one involving a transient heat load.

A large masonry wall is often used as a passive solar heating device. Consider a 300 mm thick wall composed of concrete (k = 1.8 W/m.°C, ρ = 2300 kg/m^3, C_p = 837 J/kg.°C). Since a solid model is required to study transient thermal problems, use a small rectangular section through the wall (50mm X 50mm X 300mm as shown). Since we consider flux through the wall only, the long sides of the model are insulated. The wall is initially at 20°C throughout. At time t = 0, the wall is suddenly and simultaneously exposed on the outer surface to (a) bright sunlight which adds an incident heat load of 500 W/m^2 and (b) cold air at -15°C through a convection coefficient of 10 W/m^2.°C. The inside face of the wall is exposed to a constant air temperature of 20°C through a convection coefficient of 10 W/m^2.°C. Ignore radiation from both inside and outside surfaces.

Explore this scenario as follows (you can collect most of this data by putting measures on the inside and outside surfaces, say at the midpoint):

i) For the given conditions, what is the steady state temperature distribution through the wall? What is the steady state total heat transfer into the room as a fraction of the incident solar heating? Can you verify this result analytically?

ii) For a transient analysis, plot the temperatures of the inside and outside surfaces vs time for a period of 10 hours. Is this long enough for the wall to reach thermal equilibrium? If you let the transient go long enough, does the steady state heat flux into the room agree with the value you got in part (i)?

iii) Consider a more realistic solar heating scenario where the incident heat load is a function of time. The following function should yield (does it?) the same amount of heat energy over a 10 hour period as the constant load of 500 W/m² (0.5 mW/mm²)

$$Q(t) = t*(8.33 \times 10^{-5} - 2.315 \times 10^{-9} t)$$

where Q = heat load (mW/mm²)
         t = time (seconds)

For this variable heating function, plot the temperatures on the inside and outside walls vs time for the 10 hour cycle. Create measures for the heat flux through the wall at inner and outer surfaces and plot these measures vs time. Comments?

---

10. A problem that illustrates the difference between an applied temperature and an applied heat load.

Create a flat plate with holes using the dimensions shown at the right. Units for this model are *mmNs*. The plate is 25mm thick and is made of copper. Examine the following model variations using the **3D model** geometry.

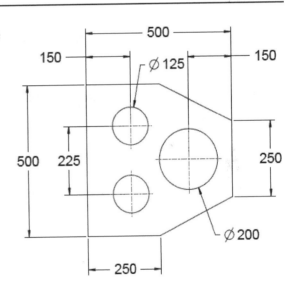

Model I (steady state)

Investigate the following thermal loading:
  All hole surfaces are maintained at a prescribed temperature of 200°C, while the left and right faces are subject to convection with h = 15.0 mW/mm².°C, T_B = 10°C. All other surfaces are insulated.
  • Show and comment on the convergence behavior of the solution.
  • Find the maximum temperature that occurs on the left and right faces. Compare the temperature distributions on the vertical edges on the front and back of the part.
  • Plot the heat flux **normal to the surface** through the left and right faces (use data from the vertical edges). Use measures at the end faces to determine the **total heat transfer out of the part**. You will need this number for the next model.

Model II (steady state)

Replace the prescribed temperatures on the holes in Model I with a **total direct heat load** equivalent to the total heat load coming out the left and right faces as computed in Model I. Divide the total load *equally* among the three holes.

- Using the measures defined on the end faces, does the total heat coming out of the model match the heat input?
- What happens to the temperatures along the edges of the holes? What and where is the maximum temperature in the plate?
- Comment on the differences in the results between Model I (specified hole temperature) and Model II (specified heat load on holes). Both involve the same energy transfer through the plate. (Hint: show the temperature fringes side by side using the same legend.)

Model III (transient)

Suppose the plate starts at an initial temperature of $10°C$ throughout. At time $t = 0$ s, the constant heat load used in Model II is suddenly applied. The convection conditions on each end are the same as before. Consider the following:

- How long does it take to reach steady state (approximately!). Does this SS temperature distribution match that of Model II?
- Suppose you want to examine a transient version of Model I (prescribed temperature on the hole surfaces). This analysis is not allowed directly by the software since the Boundary Condition and Initial Condition don't match on the holes. How could you "fake" a transient analysis with a specified temperature condition on the holes? (Hint: This involves a convective boundary condition with a physically unrealistic but mathematically valid setting!) Does this match the solution of Model I at SS?

---

11. Download the part *manifold.prt* (shown at right) from the SDC web site. This is concept model for a (hypothetical) exhaust manifold for a 4 cylinder engine. The material is steel. It is desired to determine the stress state in this body under the following steady operating conditions (all occurring simultaneously):

a) an exhaust gas pressure 50 kPa on all interior surfaces of the manifold;
b) hot gas at $200°C$ enters the manifold pipes through the four inlets at the top (forced convection);
c) all *exterior* surfaces exposed to ambient air at $25°C$ (free convection);
d) inlet flange (at the top) facing surface is at $80°C$;
e) outlet flange (at the bottom) facing surface is at $30°C$;

f) reference temperature for the manifold is 20°C.

You can assume that the inlet and outlet flanges are attached to structures at the same temperature as each flange and undergoing the same thermal expansion. Consult whatever references you like to determine the estimated convective heat transfer coefficients.

Perform a steady state thermal analysis to find the temperature distribution in the manifold. What are the maximum and minimum temperatures and where do they occur? Then in STRUCTURE, create a thermal load (MEC/T) that references the thermal design study results. Set this problem up using multiple load sets, and compare the relative magnitude of stresses and deformations for the pressure and thermal loading. Comment on the validity of these results.

HINT: For early testing of this model, keep the allowed number of passes low, or set a large convergence tolerance (20%). When you are happy with how the model has been set up, increase the allowed number of passes and decrease the convergence tolerance. Do not go below 5%, or the runs will take much too long.

IMPORTANT HINT: Remember that you have to use the same mesh for both thermal and structural analysis. You might like to do some preliminary runs with STRUCTURE to test out the mesh geometry for convergence and validity of the stress analysis before you perform the thermal analysis.

---

12. Download the part *valve.prt* (shown at right) from the SDC web site. This is the body for a hypothetical gate valve. The material is steel. The top of the valve (called the "bonnet") that contains the cover for the gate actuating mechanism is not shown and is of negligible weight. It is desired to determine the stress state in this valve body under the following steady operating conditions (all occurring simultaneously):

a) internal pressure 350 kPa (gage) on all interior surfaces of the pipe and valve body;
b) an upward force on the top flange (assume uniformly distributed) due to the pressure on the missing bonnet;
c) hot gas at 240°C passing through the valve which yields convection coefficients of 10 W/m².°C on the inside surface of the pipe entering the valve and 7.5 W/m².°C on all other *interior* surfaces of the valve body;

d) all *exterior* body surfaces exposed to ambient air at 25 °C through a convection coefficient of 5 W/m^2.°C;
e) all flange facing surfaces are insulated;
f) reference temperature for the valve is 20°C;
g) weight due to gravity.

We can assume that the flange on the pipe is allowed to deform radially and axially, but because of symmetry, will not rotate around the pipe axis. The flange on the top of the body is essentially unrestrained due to the light-weight bonnet that attaches there. The base is constructed to resist vertical motion only.

Perform a steady state thermal analysis to find the temperature distribution in the body. What are the maximum and minimum temperatures and where do they occur? Then in STRUCTURE, create a thermal load (MEC/T) that references the thermal design study results.

Set this problem up using multiple load sets (pressure, gravity, bonnet force, and thermal load), and compare the relative magnitude of stresses and deformations caused by the four loads. Comment on the validity of these results, and the way the model was set up.

HINT: For early testing of this model, keep the allowed number of passes low, or set a large convergence tolerance (20%). When you are happy with how the model has been set up, increase the allowed number of passes and decrease the convergence tolerance. Do not go below 5%, or the runs will take much too long.

IMPORTANT: Remember that you have to use the same mesh for both thermal and structural analysis. You might like to do some preliminary runs with STRUCTURE to test out the mesh geometry for convergence and validity of the stress analysis before you perform the thermal analysis.